CRYPTOCURRENCY FOR BEGINNERS MADE EASY

A NON-TECHNICAL GUIDE TO UNLOCK THE BASICS OF CRYPTO, INVEST LIKE A PRO, AND BUILD WEALTH QUICKLY WITH TAX-FREE STRATEGIES

MODERNMIND PUBLICATIONS

CONTENTS

Disclaimer:

The information in this book is for educational purposes only and should not be considered financial or investment advice. Cryptocurrency markets are highly volatile, and all investments carry inherent risks. Readers should conduct their own research and consult a qualified financial professional before making any investment decisions. The author and publisher are not responsible for any financial losses incurred as a result of using the information provided. Past performance does not guarantee future results, and investing always involves uncertainty.

INTRODUCTION

Remember January 2021? Bitcoin hit $40,000 for the first time. The news was everywhere. Your cousin who never talked about money suddenly became a crypto "expert". Your coworker quit his job to trade full-time. Your mom asked if she should buy "some of that internet money." FOMO was real.

Bitcoin kept climbing to nearly $65,000 by April. Ethereum surged. Then came the doggy coins, the ape JPEGs, and tokens named after food. People made life-changing money overnight. College kids became millionaires. Regular folks quit their jobs.

"I missed the boat," you thought.

Then May 2021 happened. Crash. Bitcoin fell over 50%. Ethereum tumbled harder. Those meme coins? Decimated. The crypto bros went quiet. News headlines shifted from "Bitcoin Millionaire" stories to "Crypto Bubble Bursts." People who bought at the top lost their savings. The guy who quit his job? He was sending out resumes again.

"Thank god I didn't jump in," you thought.

Fast forward to 2022. More pain. More crashes. The collapse of supposed "blue chip" crypto companies. FTX imploded. Luna vanished. Celsius froze. People lost millions. The crypto winter was harsh and seemed endless.

"Total scam," many said.

But then something funny happened in 2023. Bitcoin didn't die. Neither did Ethereum. The technology kept developing. The builders kept building. Away from the spotlight, the noise, and the hype, the foundations got stronger.

Then came 2024. Donald Trump's win in the U.S. election in November, coupled with his public support of the crypto space, sparked another wave. Bitcoin smashed through its previous record. Ethereum took off yet again (although not to a new all-time high). Your cousin started texting you about crypto again.

"Did I miss the boat... again?" you wondered.

Now, it's March 2025, and you're confused yet again as the market seemingly crashes, slightly recovers, then drops even further. So, what's really going on here?

Here's the truth: this is normal in investment markets. The S&P 500 drops, then recovers, then drops again. Maybe not as drastically as cryptos do, but, there is no scam and no boat to miss. That drop is what makes room for the recovery and the new gains. If the market never dropped and always went up, every day would be Groundhog's day and crypto investing wouldn't be any fun. It also wouldn't exist.

I've learned that it doesn't matter if prices are up or down today. Knowing what's really going on underneath the price charts is more important. Cryptocurrency isn't a one-time opportunity that's come and gone. It has real value and it's here to stay.

After seeing what's happened over the last several years, I've never been more excited about crypto. I get giddy now when the crypto market drops and inexperienced investors with a weak stomach panic sell all their assets. Because I know it's going to surge again, and that's my opportunity to scoop up more at a discount.

What makes this book different is I've been where you are. I've made the mistakes. I've felt the sting of losses. And I found a better path forward that doesn't require luck or timing the market.

As the founder of ModernMind Publications and an engineer by training, I've spent years immersed in the tech and cryptocurrency spaces researching, investing, and helping teach complex technical topics in ways that are clear, practical, and accessible for everyone. This book distills what I've learned into clear, practical guidance for anyone wanting to build real understanding in this space.

This isn't another flashy crypto book promising "the hottest altcoins of 2025" or "get 1,000,000% returns in 6 months." Those books exist. You might want them. But they won't teach you lasting success. You might get lucky once, then lose everything, or quit crypto because it seems "too volatile." This book is about teaching you how to invest intelligently over the long term so you can *actually* get to where you want to go.

Many people avoid cryptocurrency because it feels complex and filled with too much technical stuff that's hard to understand. The crypto market and its investors have their own language with terms that also keep people away. How do you sift through the sea of acronyms like HODL and DYOR (you'll learn what these mean) to even understand what investments to make? This book simplifies everything. You'll not only learn how to invest smartly but also explore tax-efficient strategies to grow and protect your wealth.

So if you've ever thought, "It's too late for me" or "I want to, but I'm just going to lose all my money," – I wrote this book specifically for you.

Ready for the real deal? This book is your guide to learning and exploring the intricacies of cryptocurrency investing without it needing to be intimidating or scary. Whether you're new to investing or curious about blockchain technology, you'll gain the knowledge and tools needed to make sound investment decisions. I consolidated everything I've learned into this book for you so that you don't have to make the same expensive mistakes I made or tell your spouse that you lost your kid's college fund.

If you're new to cryptocurrency investing, the best approach is to start small, learn the ropes, and develop a strategy. Everything will be broken down into clear steps, with each chapter building on the last to give you a solid foundation of knowledge you can act on immediately.

You'll start with the fundamentals of cryptocurrency and blockchain technology, then move on to major coins like Bitcoin, Ethereum, and other altcoins that are making the news. You'll learn how to securely store assets in the right wallet, choose an exchange to trade on, and develop strategies that fit your goals. We'll discuss how to evaluate coins and the technology behind them to make smart investments that have the potential to reap huge returns over the long term. I've even thrown some ChatGPT prompts in there to make it easy for you. Key topics like trends, tax basics, and staying updated in this industry are all covered. We'll also discuss how to legally invest in crypto without having to pay taxes on your gains...EVER.

By the end, you'll have a solid foundation in cryptocurrency investing, a clear understanding of how the technology works, and the strategies needed to make smart investments that get you closer to your financial goals. The future of finance and investing is developing right in front of our eyes and you don't want to get left behind.

Let's get started!

1

CRYPTOCURRENCY ORIGINS AND BLOCKCHAIN BASICS

ONCE UPON A TIME, BACK IN 2010, A PROGRAMMER NAMED LASZLO HANYECZ bought two pizzas. Normally, nobody would care if someone buys a couple of pizzas, right? But Hanyecz bought these pizzas with 10,000 Bitcoins.

Today, those same Bitcoins would be worth about $850,000,000. Not a typo.

It wasn't a bad trade back then—Bitcoin wasn't worth much, and no one really knew what it would become. Laszlo doesn't regret it either. "I think it's great that I got to be part of the early history of Bitcoin," he said later. That's the thing about history. You never know when you're making it.

That pizza purchase was the first time Bitcoin was used to buy something in the real world. A moment that proved digital money could work. A quirky experiment that set the stage for something bigger.

So, what exactly is cryptocurrency? Why does it matter? And how did a digital coin created by an anonymous coder turn into a global financial force?

It all starts with Bitcoin.

Where It All Began

Bitcoin wasn't just a random idea. It was a reaction. A response to a financial system that had let millions down.

Let's go back to 2008 when the U.S. housing market collapsed. Banks had been handing out risky loans like candy at a parade because that meant they could make more money. People took on debt they couldn't afford. When the bubble popped, everything crashed. Stocks tanked. Banks shut down. Governments scrambled to fix the mess—using taxpayer money to bail out the same institutions that caused it.

People were angry. The financial system felt broken.

Then, in October of that same year, a whitepaper titled *"Bitcoin: A Peer-to-Peer Electronic Cash System"* was published online. It was written by someone under the name Satoshi Nakamoto. Was Satoshi Nakamoto a person? A group? Nobody knew.

The idea? A new kind of money. A kind that couldn't be controlled by governments. A system where people could send and receive payments. But there was no need for a bank or any other intermediary to be involved.

Decentralization, transparency and security are the core principles driving it. Here's what that means in more detail:

- **Decentralization.** You control your money. With cryptocurrency, you don't need permission from a bank to access or transfer your funds. No one can freeze your account or tell you how to spend your money.
- **Transparency.** You can see everything. Every cryptocurrency transaction is recorded on a public ledger. You can verify transactions yourself instead of trusting banks to keep accurate records.
- **Security.** Transactions can't be altered. Built on a new kind of technology called blockchain, once transactions are verified, changing them is practically impossible.

Additionally, some coins are inflation resistant. Governments can print unlimited amounts of traditional money in an attempt to stimulate their economy. While it may be helpful in the short term, this dilutes the currency's value over time. In contrast, some cryptos (including Bitcoin) have a fixed supply. This protects it against inflation because no one can create more.

Bitcoin officially launched on January 3rd, 2009 when the first-ever block of Bitcoin was mined, called the Genesis Block. Hidden inside it was a message:

"The Times 03/Jan/2009 Chancellor on brink of second bailout for banks."

This wasn't just a timestamp. It was a statement. A middle finger to a financial system that had just collapsed under its own weight.

To get a deeper understanding of how it all works, let's break down the tech that cryptocurrency is built on.

The Tech Behind Bitcoin

Underneath every cryptocurrency is blockchain technology. At its core, a blockchain is exactly how it sounds: a chain of blocks. This chain of blocks functions as a distributed ledger, where each block in the chain is essentially a collection of records (e.g. transactions).

Each block has three key parts:

- A timestamp (when the transactions happened)
- The transaction data (who sent what to whom)
- A unique cryptographic hash (think of it as a digital fingerprint because no two blocks have the same one)

Additionally, each block stores the hash of the previous block. This links them together in a documented sequence that forms a chain. If someone tried to change a past block, they'd also have to re-calculate and re-validate every block that comes after it. This makes tampering extremely difficult and helps keep the blockchain secure.

Blockchain operates on a peer-to-peer (P2P) network, where multiple participants (called nodes) validate and store data. Put simply, several people have to agree that a block's data is valid before it can be added to the chain. This process is called a **consensus mechanism**, and every blockchain uses one. This distributed validation is what makes cryptocurrency decentralized. No single person or entity controls the blockchain or decides which transactions are allowed.

The consensus mechanism leveraged by the Bitcoin blockchain is called Proof of Work (PoW). To get a better understanding of how it works, let's cover it in more detail.

The Proof of Work (PoW) Consensus Mechanism

In Proof of Work, miners use powerful computers to compete in solving a complex cryptographic puzzle. This process is known as **hashing** and involves finding a valid hash (a unique cryptographic hash) for a new block.

Before miners even begin their work, Bitcoin transactions are broadcast to the network and are first validated by full nodes. Full nodes check transactions against Bitcoin's consensus rules (e.g., ensuring the sender has enough funds and that signatures are valid).

Once validated, these transactions go into the **mempool** (a waiting area for pending transactions). Miners pull transactions from the mempool and construct a candidate block. They typically prioritize transactions with higher fees because those generate bigger rewards.

To validate the candidate block, miners must repeatedly try different hash values until they find one that meets the network's "difficulty target". Because every block is linked to the previous block's hash, the difficulty of solving the hash or difficulty target increases as the network grows. This hashing process requires significant computational power due to the immense number of possible hash combinations. It is very energy-intensive.

Once a miner successfully finds a valid hash, they broadcast the newly mined block to the network. Other nodes then verify the block and its transactions (checking that they follow consensus rules and that the PoW is valid).

If the block is valid, it gets added to the blockchain, and the miner receives the block reward (newly minted cryptocurrency) along with transaction fees paid by users.

Pros: Highly secure and decentralized.

Cons: High energy consumption and slow transaction speeds.

How Blockchain Works (PoW)

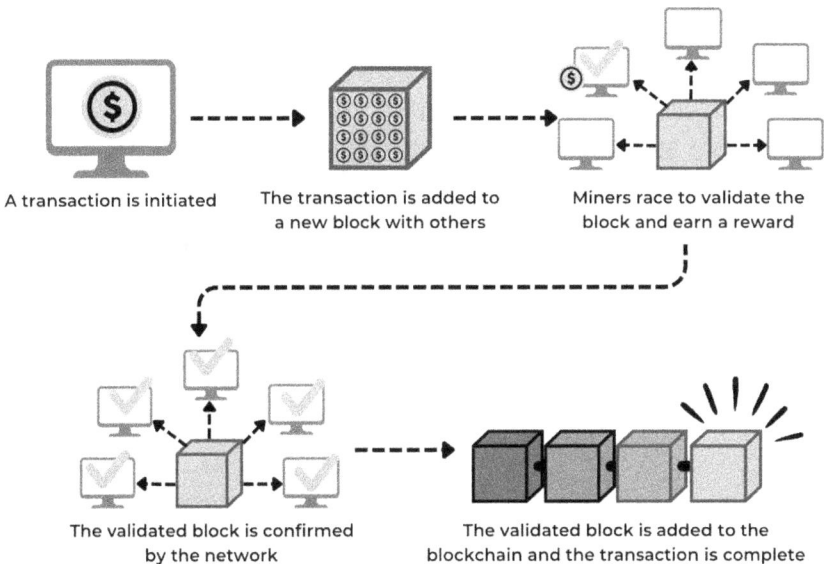

A transaction is initiated   The transaction is added to   Miners race to validate the
a new block with others   block and earn a reward

The validated block is confirmed   The validated block is added to the
by the network   blockchain and the transaction is complete

Proof of Work is not the only consensus mechanism that exists. Other blockchains utilize different consensus mechanisms. like Proof of Stake (Pos), Delegated Proof of Stake (DPoS), Proof of Authority (PoA), and Proof of History (PoH), with Proof of Work and Proof of Stake being the most common. The underlying concepts are the same for each consensus mechanism, however, in that multiple nodes have to validate transactions before they can be added to the blockchain. In later chapters, we will discuss other consensus mechanisms as we explore other blockchains.

Bitcoin's Scarcity and Halving Events

Approximately every four years, Bitcoin undergoes a halving event. In a having event, the mining reward is cut in half. This means that fewer new Bitcoins enter circulation. Historical data shows that halvings often trigger significant price increases due to reduced supply. The last halving that occurred on April 20, 2024, reduced mining rewards from 6.25 to 3.125 BTC per block. The next halving is expected around April 17, 2028, although the date isn't set in stone yet.

Bitcoin's Lasting Influence

Blockchain technology has matured a lot since the introduction of Bitcoin. While Bitcoin is primarily seen as a store of value, other blockchains have been built that have introduced a lot of technological innovation to the financial and business world. We will discuss that at length through the book.

So, if blockchain has advanced so much since Bitcoin was introduced, why is Bitcoin still so valuable? Bitcoin maintains its relevance because of its fixed supply of 21 million coins, with it often being referred to as "digital gold". Over the years, its market dominance has influenced the entire cryptocurrency ecosystem. When Bitcoin's price rises, altcoin (alternative cryptocurrencies) prices typically rise as well.

In the coming chapters, we will discuss other cryptocurrencies, starting with Ethereum and then exploring a multitude of others. In order to have a deeper understanding of why the crypto market has evolved the way it has, it's important to know what challenges blockchain technology is up against.

Challenges & Limitations of Blockchain

Here are some of the hurdles the crypto industry is working to overcome:

1. Scalability Issues: The first blockchains created, Bitcoin and Ethereum, can only handle a handful of transactions each second. When you consider that Visa and Mastercard can process thousands in that time, this has become a bigger problem as crypto adoption has grown.

2. Regulatory Uncertainty: Governments worldwide are still defining legal frameworks for blockchain and cryptocurrencies. Unclear regulations can slow down adoption and create risks for businesses.

3. Energy Consumption: PoW blockchains require high electricity usage to validate transactions.

4. Adoption and Integration Barriers: Many industries lack the technical expertise to implement blockchain solutions. Businesses must weigh costs, security, and regulatory factors before adopting blockchain.

The cryptocurrency community also has its own set of technical terms, abbreviations, and slang. For beginners, this language can seem overwhelming. But you need to speak this language so that you can:

- Recognizing fraudulent schemes and avoid scams.
- Make informed investment decisions.
- Engage in crypto discussions with experts and traders.
- Understand security practices so you can protect your investments.

How to Speak Crypto

This section is your crash course in crypto lingo. Once you get comfortable with these terms, you'll hold your own in crypto conversations, spot good opportunities from bad ones, and keep your digital money safe.

Centralized exchanges (CEXs): Platforms like Coinbase and Binance that function similarly to stock exchanges, matching buyers with sellers. You can deposit regular currency here to buy and sell crypto. They're user-friendly but require trusting the company with your funds.

Decentralized exchanges: Often called DEXs, these platforms allow direct trading between users without a middleman. They offer more control but are generally more complex for beginners.

We'll cover exchanges in more detail in Chapter 5.

Types of Wallets

Hot wallets: Digital wallets connected to the internet. Convenient for frequent transactions but less secure. Examples include MetaMask or TrustWallet.

Cold wallets: Devices or software that keep your cryptocurrency stored in a place that is disconnected from the internet. More secure but less convenient for daily use. Popular options are Ledger or Trezor.

Security Terms

Public key: Your cryptocurrency address, similar to an account number. This is safe to share with others.

Private key: The password that gives access to your cryptocurrency. Never share this with anyone.

Seed phrase: A series of 12-24 words that can restore access to your wallet if your device is lost or damaged.

Wallets and Security will be covered in detail in Chapter 4.

Crypto Community Slang

- **HODL:** Hold onto your cryptocurrency long-term, regardless of price fluctuations. Originally a typo of "hold" that became standard terminology.
- **FOMO:** Fear Of Missing Out—buying cryptocurrency because prices are rising and you don't want to miss potential profits.
- **Whale:** Someone who owns enough cryptocurrency to potentially influence market prices with their transactions.
- **DYOR:** Do Your Own Research—a reminder to do your homework before investing.
- **Pump and Dump:** A scheme where people artificially inflate a cryptocurrency's price before selling their holdings, causing its value to drop rapidly.

Common Crypto Abbreviations

- **BTC:** Bitcoin
- **ETH:** Ethereum
- **NFT:** Non-Fungible Token (unique digital assets)
- **DeFi:** Decentralized Finance (financial applications without traditional intermediaries)

- **ATH**: All-Time High (the highest price an asset has ever reached)

Crypto Transaction: Real-World Example

Let's see how this works in practice:

1. Casey buys Bitcoin on Binance. They log in, purchase 0.01 BTC, which remains in their exchange account.

2. Casey moves Bitcoin to their Trust Wallet. They copy their wallet address from Trust Wallet, enter it into Binance's withdrawal form, and complete the transfer.

3. Drew wants Bitcoin, shares their address. Casey's friend Drew sends their public key.

4. Casey sends the Bitcoin. They open Trust Wallet, select send, enter Drew's address and the amount (0.01 BTC), and confirm. The transaction includes a small network fee.

5. Bitcoin miners process the transaction. The transaction is verified and added to the blockchain.

6. Drew receives the Bitcoin. The 0.01 BTC appears in Drew's wallet.

From the early days of Bitcoin to today's trillion-dollar industry, these basic transaction principles remain the same.

In the next chapter, we'll take a deeper look at how Ethereum builds on Bitcoin's foundation and why it has become one of the most influential platforms in the crypto space.

2

ETHEREUM: A NEW VISION FOR BLOCKCHAIN

BITCOIN INTRODUCED THE WORLD TO DECENTRALIZED DIGITAL MONEY, BUT ONE man's vision took blockchain technology in a completely new direction.

In 2013, a young 19-year-old Vitalik Buterin had a revelation. Bitcoin was amazing, but limited.

"What if blockchain could do more than just send money?" he wondered.

Vitalik imagined a full-blown computer running on the blockchain. The idea was so promising that Vitalik attracted brilliant minds who believed in his vision to work on his project. Gavin Wood, Charles Hoskinson, Joseph Lubin, and others joined forces to turn the concept into reality. And, thus, Ethereum was born.

They raised funds through one of the first major cryptocurrency crowdsales in 2014. Early supporters bought Ether (Ethereum's native token) with Bitcoin, raising about $18 million worth of Bitcoin.

Ethereum finally launched on July 30, 2015. Its blockchain supports a wide range of functions that enable more complex interactions beyond simple transactions. This flexibility has made Ethereum one of the most widely adopted cryptocurrencies because it serves as the foundation for many innovations in the blockchain space.

The secret sauce? **Smart contracts.**

Smart contracts completely changed the game. But before they could be implemented on a wider scale, the Ethereum team realized that the blockchain needed a major upgrade.

The Evolution of Ethereum's Infrastructure

As Ethereum's popularity grew, so did concerns about scalability and energy consumption. Just like Bitcoin, Ethereum started out using a Proof-of-work consensus mechanism. Remember, the PoW mining process is slow, energy-intensive, and expensive.

Some of that energy comes from fossil fuels, which contributes to carbon emissions and raises serious sustainability concerns. The consequences of mining cryptocurrency this way extend beyond global energy consumption. They've affected local communities, ecosystems, and economies in undesirable ways.

Some mining operations in Kazakhstan and Russia rely on coal-fired power plants that increase carbon emissions. In regions like South America, illegal mining has led to deforestation and habitat destruction.

Because of all of this, the Ethereum team decided that a different solution was necessary.

The Ethereum 2.0 Fork: From PoW to PoS

Ethereum made waves when it switched to a more energy-efficient consensus mechanism, called Proof of Stake, with "The Merge" in 2022.

Proof of Stake, or PoS, works differently than Proof of Work, but the goal is the same: to validate transactions, create new blocks, and keep the blockchain secure.

In PoS, there are no puzzles or energy-sucking computers like in PoW. Instead, **validators** take the place of miners.

To become a validator, you have to "stake" your own cryptocurrency, or lock it up as collateral. The more coins you stake, the higher your chances of being chosen to validate the next block.

Once chosen, a validator checks that all the transactions in the block are legitimate, just like a miner does in PoW. If everything looks good, they broadcast the block to the network so it can be further validated. Once validated by a certain number of nodes on the network, the block is added to the blockchain. In return, the validator earns rewards. These are usually transaction fees or newly created crypto, depending on the network.

This shift didn't fix every issue, but it was a massive step forward. Ethereum still faces challenges like high fees and limited transaction capacity. But, the Merge reduced the network's energy use by an estimated 99.5% and laid the groundwork for a more scalable future. And that's what allowed smart contracts to truly shine.

Automating Transactions and Processes With Smart Contracts

Ethereum's blockchain was built from scratch to be programmable. The team created a Turing-complete virtual machine, the Ethereum Virtual Machine (or EVM, for short).

The EVM lets developers write and run smart contracts. Smart contracts let anyone build entire *applications* that work without banks, payment processors, or corporate platforms. Not only that, these applications run by themselves. No servers to maintain. No company that might shut down your account. No payment processors skimming fees off every transaction.

The magic happens when these contracts connect to form entire ecosystems. Lending platforms where you can borrow money without a bank. Marketplaces where artists sell directly to fans without platforms taking 30%. Games where your digital sword or land parcel is truly yours—something you can sell or rent to other players for actual money. Want to sell your house? A smart contract can hold the money in escrow, verify the deed transfer, and release payment instantly. All at 3 AM on a Sunday if that's when the deal closes.

Let's see a real example to understand how this works:

A Simple Escrow Contract in Action

- Lena wants to buy Kai's digital artwork.

- She uses a smart contract as an escrow.
- Lena sends 1 ETH to the contract address.
- The contract code records: "Lena deposited 1 ETH for Kai's artwork."
- Kai sees this deposit confirmed on the blockchain.
- Kai sends the digital artwork token to Lena.
- The contract verifies receipt: "Kai sent artwork to Lena."
- The contract automatically releases the 1 ETH to Kai.

Each step is:

- Verified by several nodes on the network
- Recorded permanently
- Visible to both parties
- Automatic once all the necessary boxes are checked

In short, Ethereum took Bitcoin's core principles—decentralization, security, and transparency—and expanded them into a fully programmable blockchain.

Smart Contract Adoption Trends

Several players across various industries across the board have noticed the value of smart contracts. The tech is being used to shake up existing processes in finance, art, and other businesses:

- **DeFi (Decentralized Finance) Expansion:** Smart contracts take banks out of lending and borrowing.
- **NFT Integration:** Smart contracts handle ownership, trading, and royalty distribution for digital assets like art, music, and trading cards.
- **Corporate Adoption:** Companies are exploring smart contracts for supply chain automation, compliance tracking, and streamlining various business tasks.

Here are some specific industry applications:

1. Real Estate

Buying or selling property has always been time-consuming. Multiple middlemen, legal paperwork, verification procedures. We've all experienced first-hand how painful it can be. Smart contracts automate these transactions by:

- Automating ownership transfers once payment clears
- Cutting out banks, escrow agents, and real estate brokers
- Facilitating transparency and stopping fraud through permanent blockchain records

With smart contracts, property deals can wrap up in minutes instead of weeks, and they cost way less.

2. Supply Chain Management

Today's supply chains get complicated and often hit snags. Products get damaged, mislabeled, lost, or stolen. Smart contracts offer a clear record that tracks every stage of a product's journey from factory to sale:

- Automates order fulfillment and payments when delivery is confirmed
- Tracks shipments in real-time
- Digitally verifying product origins

Example: Walmart teamed up with IBM to track food and reduce risks of contamination.

3. Insurance

Ever filed an insurance claim only to wait forever for a response? And then they dispute your claim and refuse to pay? Traditional insurance claims often run like molasses. Smart contracts speed this up by:

- Checking claims instantly through blockchain-stored data
- Paying out automatically when policy conditions are met
- Cutting fraud risk by recording all claims on the blockchain

For instance, flight delay insurance using smart contracts can trigger automatic payouts once real-time flight data confirms a delay.

4. Healthcare

We've all seen those crime shows where hackers take over a hospital's computers, inject malware, and demand a huge ransom. "Pay in 24 hours, or patients die." Sounds extreme, but cyberattacks on hospitals happen more than you'd think.

Medical records are gold for hackers. They have everything—identities, financial details, private health data. One breach can lead to identity theft, empty bank accounts, or even fake medical treatments in someone else's name. Keeping this info safe is tough.

Smart contracts help by:

- Enabling secure storage and controlled access to patient records
- Letting patients grant or revoke access without middlemen
- Automating billing and insurance processing based on verified medical procedures

With blockchain-backed patient data privacy, healthcare providers can boost security while sharing data smoothly across hospitals and research institutions. Estonia's government has put blockchain-based medical records into its healthcare system.

5. Digital Identity

Identity theft and data breaches are growing problems. Many of us have multiple online accounts storing our personal info. Blockchain offers a decentralized way to verify identities without relying on central databases.

- Provides secure, tamper-proof identity verification
- Cuts fraud risk by letting users control their own credentials
- Gets rid of passwords by using cryptographic authentication

With blockchain-based identity systems, people can verify their identity for banking, healthcare, and online services without exposing sensitive personal data.

Microsoft's ION project shows how decentralized identity can secure logins and verifications without traditional third-party involvement.

6. Finance & Banking

Blockchain streamlines financial transactions by providing a decentralized, transparent alternative.

- Enables instant cross-border payments without third-party approval
- Cuts transaction costs by eliminating banking intermediaries that collect high fees
- Boosts security with encrypted, tamper-proof transaction records

With blockchain-based finance, individuals and businesses can send and receive money more efficiently, reducing dependence on traditional banks.

Case Study: How Blockchain Transformed Global Banking at JPMorgan Chase

JPMorgan Chase is a giant in global finance, managing over $3.2 trillion in assets and serving millions of customers worldwide. Yet even this banking titan initially viewed blockchain with deep skepticism.

In 2017, CEO Jamie Dimon famously called Bitcoin a "fraud" and threatened to fire any employee trading cryptocurrencies. "It's worse than tulip bulbs," he said, referring to the infamous 17th-century Dutch tulip bubble.

Behind the scenes, though, JPMorgan's tech team was already exploring blockchain's potential. They saw something that would take the bank's leadership longer to admit: while crypto values might fluctuate drastically, the underlying blockchain technology offered game-changing benefits for traditional banking operations.

From Skeptic to Innovator

By 2018, JPMorgan had done a complete 180. The bank launched its

Blockchain Center of Excellence, putting over 200 developers on exploring applications of the technology.

Their research spotted several key pain points in international banking:

- Cross-border payments typically took 3-5 business days to settle
- Each transaction went through multiple banks, each taking fees
- Settlement processes needed extensive reconciliation and manual verification
- Corporate clients faced significant currency conversion costs and risks
- Tracking payment status was often impossible until completion

These inefficiencies cost JPMorgan's corporate clients billions yearly in fees, delays, and currency risks.

The Birth of JPM Coin

In February 2019, JPMorgan announced JPM Coin, becoming the first major U.S. bank to create a digital currency representing fiat money. Unlike public cryptocurrencies, JPM Coin runs on Quorum, a blockchain developed by JPMorgan (later sold to ConsenSys for wider corporate use).

How JPM Coin works:

1. A JPMorgan client deposits U.S. dollars into a designated account
2. JPMorgan issues an equivalent amount of JPM Coins on the Quorum blockchain
3. Clients use these tokens for instant payments to other JPMorgan customers
4. Recipients can redeem the tokens for U.S. dollars from JPMorgan
5. The tokens are destroyed after redemption, maintaining a 1:1 peg with the dollar

This system creates a closed loop of digital value transfer within JPMorgan's institutional client network.

By 2023, JPMorgan reported processing hundreds of billions of dollars in total transactions through its various blockchain initiatives, including JPM

Coin and its Onyx (now Kinexys) platform. The system now supports multiple currencies and serves clients in more than 100 countries. Concrete benefits include:

Dramatic Speed Improvements:

- International settlements complete in under 10 minutes instead of 3–5 days
- 24/7 operation allows transactions during weekends and holidays
- Real-time payment tracking shows transaction status at each step

Cost Reduction:

- 80% decrease in transaction processing costs
- Elimination of correspondent banking fees
- Reduced currency conversion expenses
- Lower compliance and reconciliation costs

Enhanced Security:

- Immutable transaction records prevent unauthorized changes
- Encrypted transaction data protects sensitive financial information
- Smart contract automation reduces human error
- Specialized permissions ensure appropriate access control

Business Expansion:

- Ability to serve smaller markets previously unprofitable due to banking infrastructure limitations
- New financial products built on blockchain infrastructure
- Improved treasury services for multinational corporations

Mainstream Adoption and Institutional Investment

As you can see from the JPMorgan Chase example above, blockchain fits perfectly in the corporate world. Businesses always hunt for faster and cheaper solutions to problems to boost their bottom line, and blockchain technology delivers exactly that.

With this rapid growth, cryptocurrency investing has evolved from a niche market to a mainstream financial strategy. But this also explains why cryptocurrency investing remains in its infancy. These superior technological innovations have just begun to be adopted by larger institutions, with plenty of room still to grow.

Smart contracts will become essential to modern digital transactions as blockchain networks improve scalability, security, and legal recognition.

Challenges & Considerations in Smart Contract Adoption

Despite their advantages, smart contracts still face hurdles that need addressing before widespread adoption becomes feasible.

1. Code Vulnerabilities & Security Risks

Unlike traditional contracts, smart contracts can't be easily modified once deployed. Bugs or flaws in the code can lead to exploits and financial losses.

The 2025 ByBit hack was a prime example of a smart contract vulnerability that led to $1.5 billion of Ethereum being stolen by a North Korean hacking collective called The Lazarus Group.

Rigorous auditing and testing are crucial to prevent security breaches. Blockchain developers must minimize risks by testing smart contract validation thoroughly before deployment.

2. Legal Uncertainty & Regulatory Challenges

While smart contracts automate agreements, legal systems worldwide haven't fully recognized them as enforceable contracts yet.

Jurisdictions vary in their approach to smart contract legality and dispute resolution. Regulatory frameworks are still developing, requiring dialogue between developers, legal experts, and policymakers.

For smart contracts to become a mainstream legal tool, governments must establish clearer guidelines on their enforcement and compliance.

3. Integration with Traditional Systems

Many industries still rely on legacy infrastructure that may not seamlessly integrate with blockchain-based smart contracts. The dependence businesses have on these systems to keep their operations running makes switching to a new system complex, time-consuming, and very expensive.

Organizations need hybrid solutions that allow for the gradual integration of smart contract functions into their existing systems.

4. Scalability

Smart contracts can become expensive and slow to use when the underlying blockchain is congested. As more people interact with apps built on networks like Ethereum, transaction fees can spike and processing times can increase, especially during periods of high demand. It's similar to trying to stream video content on multiple devices at once with only a 50 Mbps internet connection: everything slows down when too many users compete for limited bandwidth.

Layer 2 networks (such as Arbitrum, zkSync, and Optimism) aim to solve this by processing transactions off the main blockchain and settling them later. This makes smart contract interactions faster and more affordable.

However, these scaling solutions still need broader adoption, better user interfaces, and seamless integration with existing apps before they can fully solve blockchain's scalability bottleneck.

As smart contracts get more sophisticated, they'll increasingly serve as the foundation for a more efficient, transparent, and interconnected digital economy. This shift is already taking shape through technologies like Web3, decentralized finance (DeFi), and non-fungible tokens (NFTs), all of which function through smart contracts.

Web3 and the Decentralization of Digital Ownership

Web3 is the next evolution of the internet, designed to be powered by blockchain technology and, therefore, decentralized and use-controlled. Web3 aims to reduce reliance on centralized platforms (like Google, Facebook, or Amazon) so users have more control over their data, identity, and digital assets.

Here's how Web3 compares to earlier phases of the internet:

- **Web1 (1990s–early 2000s):** Static, read-only web pages
- **Web2 (mid-2000s–present):** Interactive, social media-driven, but controlled by centralized platforms
- **Web3 (emerging):** Decentralized, blockchain-based, focused on user ownership and autonomy

Aside from smart contracts and decentralization, these are other key features of Web3:

- **User Ownership**: People can own digital assets (like cryptocurrencies, smart contracts, or domain names) without needing a third party to manage them
- **Token-Based Economy**: Many Web3 platforms use cryptocurrencies or tokens to incentivize participation and govern networks
- **Privacy & Security**: Users can interact with Web3 applications using cryptographic wallets instead of logging in with personal data

While it's still in its early stages, its impact already shows in finance, gaming, content creation, and digital identity.

What is Decentralized Finance (DeFi)?

DeFi isn't just an alternative to traditional banking. It completely rebuilds how entire financial systems work. Unlike banks like JP Morgan Chase, which integrate blockchain to boost efficiency while keeping centralized control, DeFi removes those institutions entirely.

It uses smart contracts to run financial services like lending, borrowing, and trading through code, not people. Everything is open and automated. Users don't need approval from a bank—they interact directly with the system, and things like interest rates or loan approvals are decided by rules built into the code and real-time market activity.

The Cogs of the DeFi Wheel

To understand how DeFi works in practice, it helps to look at the main parts that make up the system. These are the tools and platforms that power DeFi's open, automated approach to finance.

Decentralized Exchanges

DeFi platforms use something called automated liquidity pools to let people trade cryptocurrencies without needing a traditional buyer and seller for each trade. Instead of matching two people directly, users trade against a pool of tokens provided by other users. These pools are managed by smart contracts and automatically adjust prices based on supply and demand. This setup makes trading faster and easier, even for newer or less popular tokens.

Unlike traditional stock markets run by big financial companies, decentralized exchanges let anyone trade straight from their wallet—no sign-up or approval needed. This system, called automated market-making (AMM), keeps trades fast and liquid.

Lending & Borrowing Protocols

Services like Aave and Compound provide peer-to-peer lending. Users deposit crypto into liquidity pools and earn interest while borrowers take loans backed by collateral. Interest rates adjust dynamically based on supply and demand. Unlike traditional banks, which require credit checks and intermediaries, DeFi lending runs automatically, uses collateral, and anyone worldwide can access it.

Stablecoins & Decentralized Payments

Cryptocurrencies like MakerDAO's DAI, USDC (USD Coin), and USDT (Tether) provide decentralized alternatives to fiat currencies. Money stored in bank accounts can be frozen, restricted, or subject to inflationary policies. Stablecoins allow users to transact globally with faster settlements, lower fees, and fewer regulatory barriers.

Yield Farming & Liquidity Provision

Investors can earn money by adding their crypto to a trading pool. In return, they get interest and sometimes governance tokens that give them a say in how the platform is run. This differs from traditional savings accounts that offer fixed and often low interest rates.

Example: Yearn.Finance is a service that automates yield farming strategies, optimizing returns by moving funds between high-yield DeFi platforms. This reduces complexity and makes DeFi more accessible to passive investors.

Synthetic Assets & Derivatives

Platforms like Synthetix enable users to trade synthetic assets that mirror the value of stocks, commodities, and other assets without owning the underlying asset. This contrasts with centralized derivatives markets, where brokers and clearinghouses act as middlemen.

As you can see, DeFi offers benefits that set it apart from traditional financial systems, including high-yield opportunities, financial sovereignty, and access to new markets. It gives users full ownership of their assets. No government. No banks. None of their rules or restrictions.

DeFi also unlocks new investment opportunities where participants can engage in crypto lending, synthetic assets, and decentralized insurance products. This levels the playing field because these markets are traditionally closed to the average retail investor.

Risks in DeFi

As exciting as all this sounds, decentralization brings risks too. Users need to actively secure their assets. Because DeFi applications run on smart contract technology, the same risks apply:

- **Code Vulnerabilities:** Flaws in smart contract code can lead to security breaches and financial losses. DeFi protocols have suffered multi-million-dollar breaches, emphasizing the need for audited smart contracts and strong security
- **Regulatory Uncertainty:** Global regulators are still figuring out how to integrate DeFi into legal frameworks. Depending on which direction those regulations go, this could impact its future growth

To address these risks, users should:

- **Research Platforms & Audits:** Check for security audits from reputable firms before investing
- **Diversify Investments:** Avoid putting all funds into a single protocol to minimize risk
- **Use Hardware Wallets:** Store assets in a cold wallet for added security
- **Stay Updated on Regulations:** Monitor legal changes to avoid compliance issues
- **Beware of Impermanent Loss:** Understand how liquidity provision affects asset value over time

Where DeFi is Headed Next

DeFi is starting to catch the eye of institutional players and traditional financial firms. Here's what's happening:

- **Institutional Investment:** Hedge funds and major financial institutions are increasingly exploring DeFi for lending, borrowing, and yield strategies.
- **Adoption by Traditional Financial Institutions:** DeFi no longer operates in isolation. Many projects are bridging crypto and traditional banking, enabling seamless on-chain (on the blockchain) and off-chain (off the blockchain) transactions through regulated gateways.
- **Expansion to Other Blockchains:** While Ethereum remains the dominant DeFi hub, networks like Solana, Avalanche, and Polkadot are emerging as faster, low-cost alternatives that aim to improve scalability and reduce transaction fees (more on this in the next chapter).
- **Privacy Enhancements:** Decentralized identity (DID) systems and zero-knowledge proofs are being developed to improve security, privacy, and compliance. This addresses one of the biggest concerns in DeFi adoption

Smart contracts and the principles of decentralization, transparency, and digital ownership that drive DeFi are also fueling the popularity of one of the weirdest, wildest corners of the crypto world: non-fungible tokens (NFTs).

NFTs: Not Just Pretty Pictures

Remember collecting baseball cards? Finding that rare Ken Griffey Jr. rookie? Or, if you're from the 90s, hunting down Princess the Bear from a kiosk while your mom waited in line at the mall? NFTs bring that same buzz except everything is digital.

NFT means "Non-Fungible Token." Sounds techie, but it's just a digital proof that you own something nobody else does. The blockchain records your ownership permanently. Your stuff stays yours. Nobody can just right-click-save your ownership away.

What can be an NFT? Pretty much anything digital:

1. Gaming Stuff

NFTs flipped gaming on its head. Before, you'd buy a $20 skin in Fortnite, but Epic Games owned it. You couldn't sell it or take it to another game.

NFT games changed that. You actually OWN your stuff.

In Axie Infinity, players breed weird little monsters that they fully own. People in the Philippines quit their jobs to play this during Covid. For real.

Gods Unchained gives you cards you can trade or sell, just like physical Magic cards. Some players make their entire income trading these cards.

2. Real Estate

Buying property used to mean you needed big money upfront. NFTs smashed that barrier.

Some companies tokenize real buildings. RealT lets you buy a piece of a Detroit rental property for $50. You get your share of rent payments weekly. No bank loans, no minimum investments.

The Sandbox and Decentraland sell virtual LAND. People build casinos, art galleries, concert venues, then charge rent. Some guy bought land next to Snoop Dogg's virtual mansion for $450,000. One plot sold for $4.3 million.

3. Music & Creative Stuff

Musicians have had to give up a huge cut of their earnings for years. Spotify pays like $0.003 per stream. Labels take most of what's left.

NFTs let artists sell directly to fans. Kings of Leon dropped an album as an NFT with concert tickets built in. 3LAU made $11.6 million in a weekend selling music NFTs.

Royal lets fans buy part of a song and earn royalties when it streams. Steve Aoki said he made more from one NFT drop than from 10 years of music advances.

Concert tickets as NFTs stop scalpers and fakes. Plus, artists get paid when tickets are resold.

NFTs work for other stuff too:

- Domain names like mike.eth sell for thousands.
- Luxury brands use NFTs as digital receipts. Your Gucci bag links to an NFT proving it's real, not some Canal Street knockoff.
- The tech fixes real problems beyond just selling expensive cartoon apes.

Making Money vs. Losing Your Shirt

Here are some of the pros and cons of investing in and owning NFTs:

The Good Stuff

- **Rare = Valuable:** Some NFTs explode in value because they're scarce and culturally significant. CryptoPunk #7523 sold for $11.8 million. A digital artist called Beeple sold a JPEG for $69 million. Not saying you'll hit these numbers, but limited collections can skyrocket.

- **Money While You Sleep**: NFTs create passive income streams that regular investments don't. Rent out your virtual land in Decentraland. Get royalties every time your music NFT gets resold. Stake your NFTs in certain platforms for token rewards.
- **Something Different**: NFTs don't follow the stock market or even Bitcoin prices. When traditional markets crashed in March 2022, blue-chip NFTs held value better than expected. They zig when other investments zag.
- **Future-Proof Value**: Today's NFTs will likely work across multiple platforms tomorrow. That CryptoPunk might just become your avatar in Facebook's metaverse, or unlock exclusive access in dozens of games and websites.

The Not So Good

- **Price Swings**: NFT prices bounce around. Floor prices for popular collections can drop 50% overnight because some celebrity tweeted something dumb. The Bored Ape that sold for $400K might fetch $50K a month later.
- **What's This Worth?**: How do you price digital art or virtual real estate? No established formulas exist. Stocks have P/E ratios. Real estate has comps. NFTs have... vibes? This makes rational investing nearly impossible.
- **Planet Killer?**: Some NFTs (especially on Ethereum) burn through electricity. Before Ethereum switched to Proof-of-Stake, minting or trading an NFT could use as much energy as a household does in a week. Regulation might hit these projects harder as climate concerns grow.
- **No Quick Exits**: Unlike stocks you can sell instantly, NFTs need a buyer who wants YOUR specific thing. Selling a rare Pokemon card takes longer than selling Apple shares. Your NFT might sit unsold for months. It's best not to count on that money for rent.
- **Scammer Paradise**: Fake collections, stolen art, pump-and-dumps, and disappearing developers plague the space. One day a collection's hot, the next day the creators vanish with millions. "Rug pulls" aren't rare—they're Tuesday.

I'm not saying avoid NFTs completely. Just know what you're getting into. Research projects thoroughly. Follow communities before investing. Start small. And never, ever spend money you can't afford to lose.

The NFT world rewards patience and punishes FOMO like nothing else.

Notable NFT Projects

1. CryptoPunks & Digital Collectibles

As one of the earliest and most iconic NFT collections, CryptoPunks consists of 10,000 unique 8-bit characters. Early adopters have seen massive returns. The record price for a CryptoPunk was $23.7 million paid in 2022.

2. Bored Ape Yacht Club (BAYC) & Social Influence

A collection of 10,000 unique ape-themed NFTs that doubles as a membership club with access to exclusive events and perks. Celebrities like Snoop Dogg, Eminem, and Steph Curry own BAYC NFTs.

3. NBA Top Shot & Sports Memorabilia

A platform for collecting digital trading cards featuring NBA highlights. It has generated millions in sales.

How to Approach NFT Investing Wisely

- **Research the Project:** Look into the creators, community, and utility of an NFT before investing
- **Evaluate Market Trends:** Study social sentiment, industry adoption, and long-term potential rather than following the craze
- **Use Trusted Marketplaces:** Stick to more reputable platforms (think OpenSea or Foundation) to avoid scams
- **Understand Liquidity Risk:** NFTs may take time to resell
- **Diversify NFT Holdings:** Avoid going all-in on a single project. Diversifying within the niche with art, music, and virtual real estate helps spread out your risk.

You're Halfway There Already

Well... that was a lot. Smart contracts. DeFi protocols. Digital ownership through NFTs. With these concepts under your belt, you're already ahead of most crypto investors.

For anyone thinking about putting money into crypto, this foundation is gold. You won't be one of those people buying at the top of the market or chasing price charts and Twitter trends. Your grasp of the fundamentals gives you a lens to make sense of the bigger picture. You're already learning the foundation of how to evaluate crypto projects with high potential. You can ask better questions about technology, adoption, and purpose.

So, with that, you're ready to get into the exciting stuff. Let's venture beyond Bitcoin and Ethereum. It's time to talk altcoins.

3

ALTCOINS: PUSHING BLOCKCHAIN EVEN FURTHER

BITCOIN AND ETHEREUM GOT THE WORLD'S ATTENTION. BUT THAT ONLY COVERS THE infancy of cryptocurrency. It's a toddler now, and altcoins are the next chapter in the blockchain story. Altcoins are simply "alternative coins," aka anything that isn't Bitcoin. Technically, Ethereum is an altcoin. There are now thousands of them, many trying to provide solutions to the shortcomings of Bitcoin or Ethereum.

Why so many? Well, people saw what Bitcoin could do and thought, "We can make this better." Some altcoins speed up transactions. Others add privacy. Some focus on smart contracts that run automatically when certain conditions are met.

You'll notice most altcoins share Bitcoin's basic DNA – they use blockchain technology to create secure, digital money without banks getting involved. But they branch out in creative ways.

When you look at an altcoin, pay attention to a few key numbers. Market cap shows the total value of all coins (price × supply). Trading volume tells you how much is being bought and sold daily. And supply mechanics explain whether there's a limited amount of coins or if new ones keep getting created. These basic stats help you compare different altcoins at a glance.

CRYPTOCURRENCY FOR BEGINNERS MADE EASY

First Generation Altcoins: Bitcoin Alternatives

Litecoin: The Faster Alternative

Charlie Lee created Litecoin in 2011 with a straightforward goal: make a faster, more accessible version of Bitcoin. He called it "silver to Bitcoin's gold," which perfectly captures what he was trying to do.

Litecoin uses a different algorithm called Scrypt. What's the big deal? Bitcoin's SHA-256 algorithm needs specialized, expensive mining equipment. Scrypt was meant to be more accessible to regular people with normal computers. This doesn't matter as much now since professional miners took over Litecoin too, but it was revolutionary at the time.

The real game-changer is speed. Litecoin processes blocks every 2.5 minutes compared to Bitcoin's 10 minutes. This means your transactions go through about four times faster. For people who got frustrated waiting for Bitcoin transfers to complete, Litecoin felt like a breath of fresh air.

Even after all these years, Litecoin stays popular. It's like Bitcoin's practical cousin – less famous but extremely reliable. Many businesses accept it, and it's usually one of the first cryptocurrencies new exchanges list. Its staying power in such a competitive market says a lot.

Bitcoin Cash: The Big Block Solution

Bitcoin Cash was born from a fight. A big, messy fight about how to handle Bitcoin's growing pains.

As Bitcoin got more popular around 2017, the network slowed to a crawl. Transaction fees shot up. People waited hours or even days for transfers to go through. Something had to change, but nobody agreed on what.

One group wanted to keep Bitcoin's 1MB block size limit but make those blocks more efficient. Another group said, "Let's just make the blocks bigger!" When nobody would budge, the network split in two on August 1, 2017. This "hard fork" created Bitcoin Cash with 8MB blocks (later increased to 32MB).

The bigger blocks worked as planned. Bitcoin Cash can handle way more transactions per second than original Bitcoin. Fees stayed lower too. If you owned Bitcoin before the split, you automatically got the same amount of Bitcoin Cash – a nice bonus for those who held on.

Today, Bitcoin Cash still has fans who believe bigger blocks were the right call. It ranks among the top cryptocurrencies by market cap, though it hasn't caught up to Bitcoin. The split shows how difficult it can be to change a cryptocurrency once it gains traction. Sometimes creating a new one is easier than fixing the old one.

Smart Contract Platforms

Ethereum isn't the only blockchain that has smart contract capabilities now. Other blockchains like Solana and Cardano were built to do things that Ethereum couldn't.

Solana: Built for Speed

Solana burst onto the scene in 2020 with an ambitious claim: it could theoretically process up to 65,000 transactions per second, though real-world throughput is usually between 2,000-5,000 right now. That's still WAY faster than Ethereum's 15-30 per second. How? Through a clever trick called Proof of History.

Proof of History is a consensus mechanism that acts like a timestamp built into the blockchain. Instead of miners wasting time agreeing on when transactions happened, Solana creates a verifiable record of time passing. This small but brilliant change supercharges its Proof of Stake system.

The results speak for themselves. Transactions cost just fractions of a penny. They complete in seconds, not minutes. This kind of performance attracted developers building everything from trading apps to video games on Solana.

Of course, there's a tradeoff. Critics point out that Solana sometimes favors speed over decentralization. The network has shut down several times when traffic got too heavy. You need pretty powerful computers to run a Solana node, which limits who can participate.

Still, Solana's growth has been incredible. Its focus on raw performance attracted billions in investment. For users tired of paying high gas fees on

Ethereum, Solana feels like using the internet after years of dial-up. A lot of very experienced investors are extremely bullish on Solana.

Cardano: The Academic Approach

Cardano takes its time, and that's the point. Founded by Ethereum co-founder Charles Hoskinson, Cardano refuses to rush. Every feature gets peer-reviewed by academics and tested exhaustively before going live.

At its heart is Ouroboros, the first provably secure Proof of Stake system. What does that mean? Basically, mathematicians have formally verified its security properties under certain assumptions, though its real-world resilience is still being proven over time. Cardano split its development into five phases with scientific-sounding names like "Byron," "Shelley," and "Goguen." Each phase adds new abilities to the network.

Smart contracts came to Cardano in September 2021, years after competing platforms. This slow pace frustrated some fans, but Cardano supporters argue that safety comes first in finance. They point to other projects that launched quickly but suffered hacks and exploits.

The Cardano community stands out for its passion. They believe strongly in the project's mission to create financial tools for people without access to banks, particularly in Africa. While other crypto projects chase quick profits, Cardano plays the long game.

Polkadot: Connecting Blockchain Islands

Polkadot tackles a problem most people didn't even realize exists: blockchains can't talk to each other. Bitcoin can't send data to Ethereum. Ethereum can't communicate with Cardano. Each chain exists as its own island.

Gavin Wood, another Ethereum co-founder, created Polkadot to build bridges between these islands. The system uses a main chain (the Relay Chain) connected to up to 100 custom blockchains (Parachains). These Parachains can be built for specific purposes but can all exchange information.

Parachains get their spots through auctions. Projects lock up DOT tokens (sometimes worth millions of dollars) to secure a two-year lease. This creates intense competition for these limited slots.

Polkadot uses Nominated Proof of Stake, where regular users pick validators they trust. This system spreads out power and lets people with small amounts of DOT participate in securing the network.

The big promise? A web where value and data flow freely between blockchains. Instead of choosing between Ethereum OR Solana OR Cardano, Polkadot wants you to use them all together as one connected system.

DeFi-Focused Altcoins

Chainlink: The Bridge to Real-World Data

Smart contracts have a big blind spot: they can't access real-world information on their own. A contract might say "pay out if it rains tomorrow," but how does the blockchain know the weather?

Chainlink fixed this problem with oracles, which are trusted data feeds that connect blockchains to the outside world. These oracles bring in stock prices, sports scores, weather data, and countless other facts smart contracts need to work properly.

But who makes sure the oracles tell the truth? Chainlink uses a network of nodes that each stake LINK tokens as collateral. If they lie, they lose their stake. Multiple nodes report the same data, which helps reduce manipulation risk, though oracle-based attacks can still occur in some systems.

Today, Chainlink secures billions across the crypto economy. Insurance products, prediction markets, and lending platforms all rely on its data feeds. When you spot "Powered by Chainlink" on a crypto project, it means they're using verified real-world data to trigger their smart contracts.

What makes Chainlink so clever? It works with any blockchain. Building on Ethereum, Solana, or something completely different? Chainlink can feed your contracts the data they need. This flexibility helped LINK become one of the most valuable cryptocurrencies out there.

Uniswap: Trading Without Order Books

Uniswap completely changed how crypto trading works by establishing the first major decentralized crypto exchange. Before it showed up, you needed order books (those lists of buy and sell orders) to trade cryptocurrencies. The system worked but needed tons of active traders to function properly.

Uniswap tossed out the order book and replaced it with math. Its automated market maker model uses simple formulas and token pools to set prices automatically. Anyone can add their tokens to these pools and earn a share of the trading fees.

Here's how it works: users create liquidity pools with equal values of two tokens (like ETH and USDC). The ratio between these tokens sets the price. When someone buys ETH, the pool gets more USDC and less ETH, so ETH's price rises a bit. This simple system runs without any middlemen at all.

Uniswap shocked its early users in 2020 with a surprise gift – UNI tokens that controlled the protocol's future. This airdrop, worth thousands to some users, kicked off a trend of governance tokens across crypto. UNI holders now vote on everything from fee structures to which blockchains Uniswap supports. It's like being on the board of directors.

Uniswap's influence goes way beyond just its own success. Hundreds of projects copied its model, and "liquidity mining" (earning tokens by providing liquidity) became a huge part of crypto culture. Even the big, established exchanges now borrow features that Uniswap created first.

Privacy-Focused Altcoins

Monero: True Financial Privacy

Bitcoin isn't nearly as private as most people think. Every transaction sits on a public blockchain anyone can view. If someone links your wallet address to your identity, they can track every penny you spend.

Monero takes the totally opposite approach. Privacy comes built-in, not as some optional extra feature. The tech behind this privacy has three main parts:

- Ring signatures hide who sent the money by mixing your transaction with others.
- Stealth addresses create one-time addresses for each transaction, so nobody can connect your payments.
- And confidential transactions hide the amounts being sent.

The result? Complete privacy, period. You can't check someone's Monero balance or see their spending history. Not even if you wanted to.

This privacy has made Monero pretty controversial. Law enforcement agencies worry about its use in illegal activities. Some exchanges have kicked it off their platforms due to regulatory pressure. But Monero supporters argue that privacy is a basic human right that shouldn't be controversial at all.

Mining Monero uses a unique algorithm called RandomX, specifically designed to work well on regular computer processors. This makes mining more accessible instead of just for those who can afford specialized equipment.

Despite all the controversy, Monero keeps a loyal following of privacy advocates, consistently ranking among the top cryptocurrencies by market cap and actual day-to-day usage.

Zcash: Privacy When You Need It

Zcash gives you a choice that Bitcoin doesn't: keep your transactions private or make them public. This flexibility comes from some seriously advanced math called zk-SNARKs (Zero-Knowledge Succinct Non-Interactive Arguments of Knowledge).

In plain English, zk-SNARKs let you prove you know something without showing what that something is. With Zcash, you can prove you have enough money without revealing your actual account balance.

Zcash has two types of addresses. Transparent addresses work just like Bitcoin, where everything shows up on the blockchain. Shielded addresses hide the sender, receiver, and amount. You can move funds between these address types whenever you want.

This "selective disclosure" approach solves a real-world problem: some situations need privacy, and others require transparency. Businesses might need private transactions with suppliers but public transactions for taxes. Zcash is designed to handle both situations, though using its privacy features can require more computing resources and wallet support.

The technology behind Zcash has spread far beyond the currency itself. Many other projects have adopted or modified zk-SNARKs for their own privacy features. Even Ethereum plans to use similar technology for scaling solutions.

Utility Tokens and Real-World Applications

Filecoin: The Decentralized Storage Market

Cloud storage typically means paying companies like Amazon or Google to keep your files on their servers. Filecoin flips this model by creating a marketplace where anyone with extra hard drive space can earn money storing files for others.

Juan Benet started the project after creating IPFS (InterPlanetary File System), a protocol for distributed file storage. While IPFS handles how files move around, Filecoin adds economic incentives to make sure those files stay available over time.

The system uses two novel consensus mechanisms. Proof of Replication verifies that storage providers actually store the data they claim to. Proof of Spacetime confirms they keep storing it continuously over time. Providers stake FIL tokens as collateral, which they lose if they fail to maintain the files properly.

Compared to traditional cloud storage, Filecoin offers potentially lower costs and better censorship resistance. No single company controls the network, so files can't be taken down by government pressure or corporate policy changes.

The real question is whether Filecoin can compete with established storage giants on price and convenience. Storing data on AWS or Google Drive is simple and cheap. Filecoin needs to match that user experience while offering enough benefits to make switching worthwhile.

Basic Attention Token: Fixing Online Advertising

Online ads track your every move. They slow down websites. They eat your data plan. And despite all this, most of the money goes to the company displaying the ad instead of the content creators you love.

Basic Attention Token (BAT) and the Brave browser tackle these problems by completely rethinking how ads work online. Their approach puts you in control of your attention, which they correctly identify as the valuable resource everyone's fighting for.

Brave blocks traditional ads and trackers by default. This makes websites load faster and keeps your browsing private. But then it adds a twist: you can opt in to see Brave's own ads, which pay you BAT tokens for your attention. These ads appear as system notifications, not webpage clutter.

Publishers and content creators receive BAT when you spend time on their sites. You can also tip your favorite creators directly. The system cuts out the ad networks and data brokers that typically take most of the money.

The numbers show growing interest. Brave has millions of monthly users. Many YouTube creators, Twitch streamers, and websites have registered to receive BAT. Major brands run ad campaigns through the platform.

While BAT hasn't overthrown Google or Facebook yet, it offers a glimpse of what advertising could be: respectful of privacy, fair to creators, and actually valuable to users who choose to participate.

Stablecoins: Bridging Traditional and Crypto Finance

USDC and USDT: Digital Dollars

Bitcoin might jump or drop 10% in a day, making it great for speculation but terrible for everyday use. Stablecoins solve this problem by maintaining a steady value, usually pegged to a traditional currency like the US dollar.

USDT (Tether) pioneered this approach in 2014. USDC followed in 2018 as a more regulated alternative. Both work on a simple premise: USDT is backed by a mix of assets, including cash, cash equivalents, and other reserves—not just dollars in bank accounts. This 1:1 reserve ratio keeps the price stable at (roughly) one dollar.

Both coins play crucial roles in crypto markets. They give traders a safe place to park money during market turbulence. They enable fast transfers between exchanges. And they provide a familiar unit of account for people new to cryptocurrency.

But there are key differences. USDC, created by Circle and Coinbase, publishes monthly attestations of its reserves from major accounting firms. USDT has faced questions about its backing and settled with the New York Attorney General after an investigation into its reserves.

The impact of these stablecoins goes beyond trading. They enable fast, cheap international payments without currency conversion fees. They bring dollar stability to people in countries with weak local currencies. And they form the foundation of many lending and saving platforms in the crypto world.

DAI: The Algorithmic Approach

DAI takes a radically different approach to stability. Instead of keeping dollars in a bank, it uses excess cryptocurrency as collateral. This removes the need to trust a central issuer but introduces new complexities.

The system works through "vaults" on the MakerDAO platform. You deposit cryptocurrency worth more than the DAI you want to generate – typically at least 150% more. This over-collateralization protects against market drops. If your collateral value falls too low, the system liquidates it to maintain DAI's dollar peg.

MakerDAO, the organization behind DAI, runs as a Decentralized Autonomous Organization. MKR token holders vote on critical parameters like collateral types, required collateral ratios, and stability fees (essentially interest rates for creating DAI).

This approach has pros and cons. DAI doesn't face the regulatory risks of centralized stablecoins. Nobody can freeze your funds or censor your transactions. However, the system can become unstable during extreme market conditions, as happened during the March 2020 crypto crash when many vaults got liquidated.

Despite these challenges, DAI has largely maintained its peg over time, even after brief periods of volatility. Its success inspired many other algorithmic stablecoins, though most have proven less resilient. For people who value true decentralization, DAI offers something no centralized stablecoin can: financial stability without centralized control.

Future Trends in Altcoin Development

The altcoin space never stands still. Several clear trends show where things are heading next.

Layer 2 solutions will continue growing to solve blockchain scaling problems. As mentioned briefly in the last chapter, these systems handle transactions off the main chain and periodically settle the final results instead of making base blockchains faster (which often reduces security or decentralization). Optimistic rollups and zero-knowledge rollups lead this category, potentially increasing transaction throughput by 100× or more.

Cross-chain technology will become standard, not special. Projects like THORChain and Cosmos join Polkadot in building infrastructure that lets different blockchains communicate seamlessly. The walls between crypto ecosystems will gradually fall as users demand the ability to move assets and data between chains.

Tokenization of real-world assets is starting to grow quickly. Things like treasury bonds, real estate, and carbon credits are now being turned into digital tokens on the blockchain. This means traditional assets can be traded using crypto technology. It could open up a market potentially worth hundreds of trillions of dollars.

AI integration with blockchain has barely started but shows exciting potential. Machine learning models could help set optimal parameters for DeFi protocols, predict network congestion, or identify suspicious transaction patterns. When combined with privacy technology, AI could even analyze encrypted data without compromising user privacy.

Altcoins have transformed blockchain from a single experiment into a thriving ecosystem of specialized solutions. Bitcoin proved the core concept, but altcoins expanded what's possible and fixed critical limitations.

You now understand the major categories: Bitcoin alternatives that focus on speed and scaling, smart contract platforms that enable complex applications, privacy coins that protect financial confidentiality, utility tokens that solve real-world problems, and stablecoins that bridge traditional and crypto finance. We couldn't possibly cover all the cryptocurrencies out there in one chapter or even one book. So, from here on out, you'll have to DYOR. Later, when we reach Chapter 8 on fundamental analysis, you'll learn exactly how to evaluate which of these altcoins might make good investments. The technical foundations we've covered here will help you make sense of those investment decisions. Here's a table to summarize all that we've covered in this chapter:

ALTCOIN	WHAT IT DOES	WHY IT'S DIFFERENT
Litecoin (LTC)	A faster version of Bitcoin	Quicker transactions and easier to mine (at first)
Bitcoin Cash (BCH)	A Bitcoin spin-off	Handles more transactions with lower fees
Solana (SOL)	A fast smart contract platform	Super quick and cheap transactions
Cardano (ADA)	A careful, research-based platform	Built slowly with peer-reviewed science
Polkadot (DOT)	Connects different blockchains	Lets blockchains work together like one big network
Chainlink (LINK)	Feeds real-world data to blockchains	Helps smart contracts know things like prices or weather
Uniswap (UNI)	A place to trade crypto without a middleman	Uses math instead of order books to set prices
Monero (XMR)	Keeps your transactions private	Hides sender, receiver, and amounts by default
Zcash (ZEC)	Lets you choose privacy or transparency	You decide if a transaction is public or private
Filecoin (FIL)	Pays people to store files	Decentralized alternative to Google Drive or Dropbox
Basic Attention Token (BAT)	Pays you for watching ads	Works with the Brave browser to reward your attention
USDT (Tether)	A stable crypto tied to the dollar	Good for saving or trading without price swings
USDC	A regulated, stable digital dollar	Similar to USDT, but with more transparency
DAI	A stablecoin without a central company	Backed by crypto, not dollars in a bank

NOW THAT YOU understand the basic principles and mechanisms behind blockchain technology, Bitcoin, Ethereum, and major altcoins, where do you actually store your crypto? In the next chapter, we'll look at wallets.

4

WALLETS: STORE AND PROTECT

You can think of a cryptocurrency wallet a lot like the one your dad has carried around for the past 30 years, except it's digital, stores crypto instead of fiat money, and is probably in better condition. And there's no expired Blockbuster card, no crumpled receipts, and definitely no 3rd-grade school picture from the year your mom gave you that tragic bowl cut. Maybe they aren't that similar, actually. Because it's also a lot more difficult to lose your crypto wallet than a physical wallet. As long as you don't lose your private key.

This brings us to our next topic: how to store and protect your crypto. Keeping your cryptocurrency safe is as important as choosing which assets to invest in. That's why it's imperative we cover this topic *before* you start investing. Cryptocurrencies operate under the principle of self-custody, meaning that you are solely responsible for the security of your assets. You need to be both cautious and proactive. If you lose them, they're gone.

Types of Wallets

There are different kinds of crypto wallets and not all of them are created equal. They each have their benefits and drawbacks. Understanding the different types of wallets will help you make an informed decision on which one is the best for you and your investment goals, rather than just choosing the most convenient option and regretting it later.

Custodial vs. Self-Custody Wallets

- **Custodial wallets** are managed by third parties, such as cryptocurrency exchanges or investment platforms. They are convenient, but they also require you to trust the service provider to store and manage your assets safely. These wallets are often found on centralized exchanges (CEXs) where you trade, or within institutional investment accounts like crypto IRAs. If the platform is hacked or freezes withdrawals, you could lose access to your funds.

- **Self-custody wallets** (also called non-custodial wallets) give you full control over your private keys and funds. This means you are entirely responsible for securing them. If you lose your keys, you lose access to your assets permanently. Self-custody wallets require more effort to manage securely, but many investors prefer them for the added protection.

Hot vs. Cold Wallets

To add a layer of complexity to this, there are also hot wallets and cold wallets.

Hot wallets are cryptocurrency wallets that are connected to the internet. They're highly convenient for everyday transactions because accessing your funds is easy. These wallets come in various forms, like mobile apps, desktop software, and web-based wallets. When you trade on a centralized exchange and keep your investments held there, they are in a hot wallet.

Advantages of Hot Wallets

- **Instant Access**: You can quickly buy, sell, or trade crypto anytime.
- **User-Friendly**: Ideal for beginners due to simple interfaces.

- **Integration with Exchanges**: Many hot wallets connect directly to crypto exchanges for seamless trading.

Disadvantages of Hot Wallets

- **Higher Security Risks**: Being online makes them more vulnerable to hacking, malware, and phishing attacks.
- **Custodial Risk**: Some hot wallets (like those offered by exchanges) are custodial hot wallets that hold your private keys, meaning you don't have full control over your assets.

Best Practice: Only keep the amount of crypto in a hot wallet that you need for daily transactions. Consider cold storage for larger amounts.

Cold wallets are a good option if you're looking for something more secure because they store your cryptocurrency offline. These can either be in the form of software that relies on offline storage methods or hardware wallets that are actual physical devices. Since they are not connected to the internet, they are much more immune to hacking attempts, phishing, and malware attacks.

Why Use a Cold Wallet?

- Ideal for storing large amounts of cryptocurrency long-term.
- Significantly reduces online attack risks by keeping private keys offline.
- Protects assets from exchange failures, fraud, or government restrictions.

While cold wallets offer strong protection, they are not immune to theft if private keys are exposed or if the user's device is compromised.

Types of Cold Wallets

1. **Hardware Wallets:** Devices like Ledger Nano X and Trezor Model T store your private keys offline while allowing secure transactions when connected.

Pros: Highly secure, easy to use, supports multiple cryptocurrencies.

Cons: Requires an upfront cost ($50-$200) and should be physically protected in a fireproof safe.

2. **Paper Wallets:** Consist of a printed QR code or written private key stored on physical paper.

Pros: No exposure to online threats, highly secure if properly stored.

Cons: Risky. If lost or damaged, funds are irrecoverable.

3. **Air-Gapped Wallets:** These are software wallets installed on devices that never connect to the internet. Used by institutional investors and high-net-worth individuals for extra protection.

Best Practice: Use a hardware wallet for long-term holdings and store recovery phrases in a secure location.

Hot vs. Cold, Custodial vs. Non-Custodial: How They Overlap

Here's where it gets tricky. Every wallet is either a custodial wallet OR a non-custodial wallet, and every wallet is either a hot OR a cold wallet. But you can have wallets that are custodial AND hot wallets or non-custodial AND cold wallets. In fact, all wallets have both custodial and connectivity components. There are:

- **Custodial Hot Wallets: Connected** to the internet and you **don't** own the private key. Examples: Binance, Kraken, Coinbase
- **Non-Custodial Hot Wallets: Connected** to the internet and you **do** own the private key. Examples: MetaMask, Trust Wallet, Exodus
- **Custodial Cold Wallets: Not connected** to the internet and you **don't** own the private key (usually rare and used by large institutions). Examples: Coinbase Custody and BitGo
- **Non-Custodial Cold Wallets: Not connected** to the internet and you **do** own the private key. Can be either software or hardware. Examples: Hardware wallets like Ledger and Trezor

Exercise: Setting Up Your Wallet

Here's a step-by-step guide on how to choose your wallet and get it set up:

Step 1: Choose the Right Wallet

Based on our discussion of the different types of wallets, there are three questions you need to ask yourself when you evaluating your options:

- **Control:** How much control do you want to have over your investments?
- **Connectivity:** How accessible do I need my investments to be?
- **Security:** How secure do I want my investments to be?

Assessing your needs will help you select the best wallet type to secure your holdings. Here's a matrix that can help you evaluate:

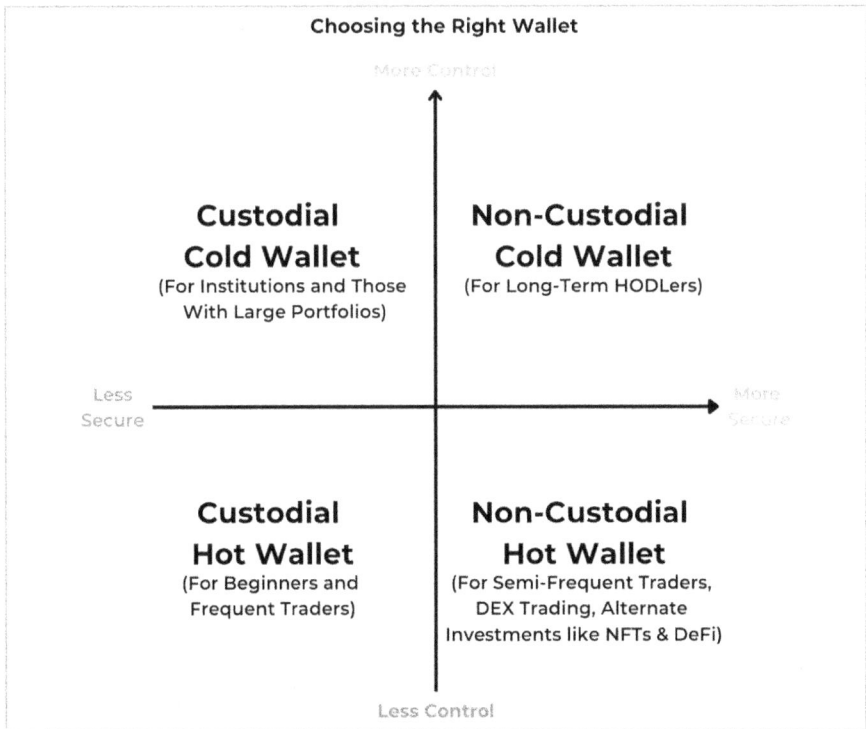

Choosing the Right Wallet

	More Control	
Custodial Cold Wallet (For Institutions and Those With Large Portfolios)		**Non-Custodial Cold Wallet** (For Long-Term HODLers)
Less Secure		More Secure
Custodial Hot Wallet (For Beginners and Frequent Traders)		**Non-Custodial Hot Wallet** (For Semi-Frequent Traders, DEX Trading, Alternate Investments like NFTs & DeFi)
	Less Control	

Remember, you don't have to choose just one. Many investors use both a hot wallet for daily transactions and a cold wallet for long-term holdings. That means you just need to choose whether you want custodial or non-custodial wallets and then select one of the available options on the market. Most investors start with a custodial hot wallet as they dip their toes into investing on a centralized exchange.

Step 2: Generating and Securing Your Seed Phrase

Once your wallet is set up, it will generate a seed phrase (recovery phrase). This is a sequence of 12 to 24 random words that acts as the master key to your funds. If you lose access to your wallet, your seed phrase allows you to recover it and all its private keys.

How to Secure Your Seed Phrase

Anyone with your seed phrase can grab your crypto. Treat it like you would a birth certificate or the antique jewelry your grandma gave you: keep it protected.

- Write it on paper and lock it somewhere safe (like a fireproof safe or safety deposit box).
- Never save it digitally (on your phone, computer, or in the cloud).
- Look into metal backups like Cryptosteel or Billfodl that are both fire and waterproof

Golden Rule: Someone asking for your seed phrase? 100% scam. Real services never ask for it. Ever.

Step 3: Testing Wallet Recovery

Lots of people set up wallets but never test recovery, just assuming their seed phrase will work when needed. But if there's any mistake or your backup is wrong, you might lose everything permanently.

You can avoid this nightmare by doing a wallet recovery test:

1. Set up a new wallet on a different device
2. Enter your seed phrase to restore it
3. Check that your funds show up correctly

If recovery works, lock that seed phrase away somewhere only you can access.

More Ways to Protect Your Wallets

Protecting your crypto goes beyond picking the right wallet. These strategies will help shield your digital money from hackers, scammers, and unexpected problems.

Securing Networks & Devices: Defending Against Cyber Threats

Your crypto is only as safe as the devices and networks you use. Hackers use malware, phishing, and other attacks to steal cryptocurrency, often targeting careless internet users.

Here are the security steps you should take:

- **Use a VPN:** It protects your internet connection when you check your crypto accounts.
- **Skip Public Wi-Fi:** Hackers love intercepting data on these networks.
- **Get Anti-Malware:** Protect against keyloggers with Wallet Guard (a free browser extension).
- **Turn On Two-Factor Authentication:** Always use an authentication app (not SMS) for extra protection.

Maintaining Privacy: Reducing Exposure

While crypto transactions aren't tied to your name, they're not completely private. Blockchain transactions can be traced, and personal details can leak if linked to your wallet.

Here are some ways to boost privacy:

- **Use a Separate Email for Crypto:** Stop scammers from connecting your personal identity to your crypto.
- **Don't Share Wallet Addresses Publicly:** Transactions can be tracked, so limit who sees your address.

- **Consider Privacy Coins:** Remember, cryptos like Monero and Zcash offer better anonymity.

Protecting Private Keys

Owning crypto in a non-custodial cold wallet means you control your private keys. If someone gets your private key, they can steal everything, and you can't get it back.

In 2023 alone, hackers stole $1.7 billion from crypto funds, showing why strong security matters.

Key Security Tips:

- NEVER share private keys or recovery phrases with anyone.
- Use strong passwords. Don't reuse passwords.
- Turn on two-factor authentication where possible.
- Back up your recovery phrase in multiple secure places.

Golden Rule: If you don't control the private keys, that crypto isn't really yours.

Regular Security Check-Ins

One of the most overlooked parts of crypto security is regular security evaluations. Just like maintaining your car or reviewing investments, periodic security checks ensure your defenses stay strong.

How to do a thorough security audit:

- **Check Your Wallet Security:** Make sure your wallet software is updated and your seed phrase is secure.
- **Look for Phishing Risks:** Check email security and avoid suspicious crypto-related links.
- **Update Everything:** Keep wallet apps, browser add-ons, and hardware wallet firmware current.
- **Keep an Eye On Account Activity:** Check your exchange login history regularly for any weird access.

Best Practice: Set a reminder every few months to check security on all your crypto accounts and devices.

Diversification: Reducing Risk with Multiple Wallets

A basic investment principle is diversification, and this applies to storing crypto too. Using just one wallet or exchange puts all your eggs in one basket. If it gets compromised, everything's gone.

Smart wallet diversification tips:

- **Use More Than One Wallet:** Keep a hot wallet for everyday use and a cold wallet for savings. Consider multiple cold wallets for large holdings.
- **Limit Exchange Balances:** Don't keep big amounts on exchanges.
- **Spread Funds Across Different Blockchains:** Holding assets on several networks reduces risks tied to technical problems or regulations.

Example: A 2023 hack on Atomic Wallet resulted in over $35 million in losses —a brutal reminder of why keeping funds in multiple wallets makes sense.

A Secure Foundation

Throughout this chapter, we've covered crucial crypto security that every investor needs to know. We've looked at different wallet types and explained why self-custody through non-custodial and cold wallets gives you the best security and control.

We've gone through basic security practices that should become automatic: strong passwords, two-factor authentication, and backing up seed phrases in multiple secure spots.

Remember: with cryptocurrency, YOU are the bank. This provides amazing financial freedom but puts security squarely on your shoulders. These measures aren't optional extras. Don't learn this stuff the hard way.

Now that you have a solid footing in securing your crypto, you're ready to find an exchange so you can actually do the thing you wanted to do when you bought this book: invest. In the next chapter, we'll guide you through finding and choosing a good exchange. We'll explore different types available and help you understand which ones fit your trading goals.

5

CRYPTOCURRENCY EXCHANGES: BUY, SELL, AND TRADE

Crypto exchanges are marketplaces (usually apps) where people buy, sell, and trade digital currencies. Similar to stock exchanges for traditional investments, these platforms form the backbone of the crypto economy and give you access to various digital assets. The big difference? Unlike the stock market, crypto never sleeps. You can trade 24/7.

In this chapter, we'll break down the essentials to help you find the right exchange and start trading with confidence.

Types of Exchanges

Finding the right exchange marks an important step for new crypto investors. With so many options offering different features, fees, available coins, and security measures, things can feel confusing at first. Some platforms cater to beginners with simple interfaces, while others offer advanced tools for experienced traders.

There are three main types of cryptocurrency exchanges, each with its own pros and cons:

1. Centralized Exchanges

Ever used apps like Robinhood or eToro for stocks? You create an account, deposit money, and start buying stocks within a day. Centralized exchanges, also knowns as CEXs, work very similarly. Users trust a third-party platform to handle transactions. These user-friendly exchanges work great for beginners.

Advantages:

- **Fiat Compatibility:** Easily buy crypto using bank transfers, credit cards, and PayPal
- **Lots of Liquidity:** More users mean faster trades with minimal price slippage
- **Customer Support:** Get help through support desks, FAQs, and chat
- **Regulated Operations:** Many CEXs follow government regulations and offer insurance

Disadvantages:

- **Centralized Control:** You don't control your private keys; assets sit on the exchange's servers. The exchange manages your funds, so you must trust them to keep things secure and allow withdrawals
- **Security Risks:** CEXs often become targets for hackers. Popular exchanges take extreme security measures, though
- **Regulatory Restrictions:** Some exchanges have regional limits and require KYC (Know Your Customer) identity checks

Popular CEXs:

- **Binance:** Offers tons of cryptocurrencies, advanced trading features, and Binance Lite for beginners
- **Coinbase:** Known for its easy interface, strong security, and great educational resources
- **Kraken:** A well-respected platform with tight security and many supported coins

2. Decentralized Exchanges

We've already talked about decentralized exchanges in our chapter 2 discussion of DeFi and our chapter 3 discussion of UniSwap, but let's revisit. Decentralized exchanges run without any central authority. Users trade cryptocurrencies directly (peer-to-peer) from external hot wallets like MetaMask. Unlike CEXs, DEXs don't hold your funds or personal data, giving greater security and privacy.

However, DEXs only support assets on the same blockchain they run on. For example, Uniswap only handles Ethereum-based tokens (ERC-20), while PancakeSwap sticks to Binance Smart Chain tokens (BEP-20). If a token isn't native to the blockchain the DEX uses, you can't trade it there without a cross-chain bridge.

Advantages:

- **Full Control:** You keep control of your private keys.
- **Privacy & Anonymity:** Most DEXs don't require identity verification.
- **No Centralized Risks:** With no central entity, hacks and regulatory shutdowns become less likely.

Disadvantages:

- **Lower Liquidity:** Some DEXs have fewer traders, leading to price slippage.
- **No Fiat On-Ramps:** You need to already own crypto to trade.
- **Limited Asset Availability:** DEXs only support tokens native to their blockchain. Cross-chain trading requires bridges. This can make things riskier and more complex.
- **Less Beginner-Friendly:** DEXs require managing private keys and smart contract interactions. Trading often involves multiple steps: connecting a wallet, setting slippage tolerance, and manually approving transactions. UniSwap and PancakeSwap are DEXs working to simplify things.

Popular DEXs:

- **Uniswap:** One of the biggest Ethereum-based DEXs for token swaps
- **PancakeSwap:** A Binance Smart Chain DEX with lower fees and faster transactions
- **dYdX:** dYdX is transitioning from a semi-decentralized model to a fully decentralized exchange with its new v4 version, offering derivatives and margin trading on its Cosmos-based chain.

3. Hybrid Exchanges

Hybrid exchanges try to combine features from both centralized and decentralized exchanges. Many offer non-custodial trading where you're in control of your private keys but provide a centralized order book for faster trades and better liquidity. This gives you CEX-like speed and convenience with greater security than traditional exchanges.

Pros:

- **Partial control over funds:** Some hybrid exchanges let you manage private keys, though others keep some custodial elements.
- **User-friendly interface:** Smoother experience than most DEXs while keeping some decentralized features.
- **Better liquidity than DEXs:** Uses order book models or centralized matching for faster trades while improving security over CEXs.

Cons:

- **Not fully decentralized:** Some hybrid exchanges still manage some user assets or use centralized infrastructure.
- **Limited adoption and liquidity risks:** Because hybrid exchanges aren't as widely used as CEXs or DEXs, liquidity and trading pairs may be limited.
- **Potential regulatory challenges:** Hybrid models might face uncertainty because they have to factor in both CEX and DEX regulations.

Examples:

- **Rhino.fi (formerly DeversiFi):** Offers privacy-focused trading on Ethereum with fast transactions

Beginner-Friendly Crypto Exchanges

For crypto newcomers, these exchanges provide simple interfaces, educational resources, and strong security:

- **Coinbase:** Best for beginners thanks to its intuitive platform and regulatory compliance.
- **Binance:** Offers Binance Lite for easy entry into trading, plus lots of learning materials.
- **Kraken:** Known for solid security and extensive coin support, perfect for those looking to grow beyond beginner levels.

You'll notice all these are centralized exchanges—they have a much easier learning curve. Each platform serves different needs, so consider your priorities (security, ease of use, fees, or available cryptocurrencies) when picking an exchange.

Key Factors and Features to Consider

Beyond choosing between CEX, DEX, or hybrid exchange, evaluate these additional factors:

1. Reputation & Security

Research the exchange's history, reviews, and past security issues. Make sure the platform uses strong encryption and two-factor authentication. Look for insurance policies protecting user funds

2. Supported Cryptocurrencies & Trading Pairs

Check if the exchange offers the coins you want to buy, sell, or trade. Some platforms provide only major coins like BTC and ETH, while others list hundreds of altcoins. Binance offers over 600 coins, but more isn't always better, especially for beginners.

3. Fees & Transaction Costs

Compare trading fees, deposit fees, and withdrawal charges. Some exchanges have maker-taker fee structures, with discounts for high-volume traders.

4. Liquidity & Market Depth

More liquidity means fast trades with minimal price fluctuation. Exchanges with low liquidity may cause higher spreads and slippage.

5. User Experience & Mobile Access

Look for platforms with easy-to-use interfaces. If you prefer trading on the go, check for a well-designed mobile app.

6. Regulation & Compliance

Some exchanges require KYC verification, while others allow anonymous trading. Verify if the exchange legally operates in your country to avoid restrictions.

7. Wallet Integration & Custodial Services

Determine if the exchange provides built-in wallets or needs external wallet integration. Some platforms offer custodial wallets, while others require self-custody.

Exchange Security Features & Best Practices

Security stands as one of the most critical factors when choosing a cryptocurrency exchange. With billions in digital assets flowing through these platforms, strong security measures must protect user funds from cyber threats, fraud, and exchange failures.

A well-secured exchange uses multiple protection layers, from account-level security like two-factor authentication to institutional-grade cold storage that keeps most assets offline. Understanding these security features helps you assess an exchange's reliability and protect your investments.

Key Security Features to Look for in an Exchange

1. Two-Factor Authentication

Two-factor authentication serves as your first defense against hackers. Instead of just using a password, 2FA requires an extra verification step, such as:

- A one-time code sent via SMS or email.
- An authentication app like Google Authenticator or Authy.
- A hardware security key (like YubiKey) for maximum protection.

Advantages:

- Stops unauthorized access even if your password gets compromised.
- Protects against phishing attacks and brute-force login attempts.
- Required on most reputable exchanges, including Binance, Kraken, and Coinbase.

Disadvantages:

- There are never any downsides to 2FA aside from the minor inconvenience of protecting your identity.

Best practice:

Always turn on 2FA for your exchange account and use an authenticator app or hardware security key when possible. SMS-based authentication can fall victim to SIM-swap attacks.

2. Cold Storage

Look for an exchange that stores the majority of their holdings in cold wallets.

Advantages:

- Reduces hacking risks by keeping most assets inaccessible from the internet.
- Prevents large-scale exchange breaches from affecting all user funds.

- Many top exchanges, including Binance, Kraken, and Coinbase, store over 90 percent of digital assets in cold wallets.

Example:

In 2019, Binance suffered a security breach, losing 7,000 BTC (~$40 million at the time). However, because most funds stayed in cold storage, users remained unaffected, and Binance reimbursed losses through its Secure Asset Fund for Users (SAFU).

3. Multi-Signature Wallets

Multi-signature (multi-sig) wallets need multiple approvals before executing a transaction. Most standard wallets only require a single private key to move funds. Multi-sig wallets demand two or more signatures from designated parties.

Advantages:

- Prevents a single point of failure because no individual can move funds alone.
- Adds extra security for large transactions and institutional funds.
- Helps reduce risks from insider threats or compromised keys.

Example:

- BitGo, a digital asset custody provider, implements multi-signature cold wallets to enhance security for institutional clients.
- Many crypto hedge funds and large investment firms use multi-sig wallets to protect high-value transactions.

Best practice: If your exchange offers multi-sig security, use it, especially for large withdrawals.

4. Insurance Coverage & Fund Segregation

Many leading exchanges provide insurance policies protecting user funds against hacking, fraud, and operational failures. Additionally, fund segregation ensures that:

- User assets are kept separate from company operating funds.
- Exchanges cannot use customer funds for their own expenses or trading.
- In the event of insolvency, user funds are more likely to be recoverable.

Examples:

- Coinbase carries crime insurance for digital assets held in its hot wallets. This does not cover losses from user-side breaches like phishing or stolen credentials.
- Gemini Exchange is insured by Aon, providing additional user protection.

Best practice: Check if an exchange offers insurance and confirm how they manage user funds during financial troubles.

5. Transparency & Security Audits

Reputable exchanges continuously test vulnerabilities and improve defenses through security audits. Transparent platforms often:

- Publish audit reports detailing how they secure funds.
- Implement proof-of-reserves systems showing they hold enough assets to cover customer balances.
- Provide detailed records of past breaches and improvements.

Best practice: Choose an exchange that publicly shares its security practices, audits, and fund management policies.

Even with strong security practices in place, some platforms have suffered major breaches in the past. Understanding these incidents can help you spot red flags and choose safer exchanges.

Past Exchange Hacks

Some notable exchange breaches include:

1. The Mt. Gox Collapse (2014): The Largest Bitcoin Hack in History

Once the largest Bitcoin exchange, Mt. Gox lost 850,000 BTC (worth $450 million at the time) due to security flaws. Poor internal controls and lack of cold storage left funds vulnerable to long-term, unnoticed theft. The exchange collapsed, leaving thousands of users with no options to recover their funds.

2. The Bitfinex Hack (2016): Multi-Signature Exploit

Hackers exploited a vulnerability in Bitfinex's multi-signature wallet implementation with BitGo, leading to the loss of 120,000 BTC.

3. The Coincheck Hack (2018): $530 Million Stolen

Coincheck lost 523 million NEM tokens due to weak security measures. Unlike other major exchanges, Coincheck stored large amounts of funds in hot wallets. They were an easy target.

How to Protect Yourself When Using an Exchange

Follow these best practices to keep your investments safe:

- **Enable Two-Factor Authentication**: Just as with protecting private keys, always activate 2FA for login, withdrawals, and account changes.
- **Use Strong Passwords**: Avoid reusing passwords across multiple platforms.
- **Withdraw Long-Term Holdings to a Cold Wallet**: Don't leave large amounts of crypto on an exchange. Instead, store them in a hardware wallet like the ones we discussed in chapter 4. Check to see if the exchange you use stores crypto in multi-sig cold wallets. This should be public information.
- **Verify Withdrawal Addresses**: Always double-check addresses before sending funds.
- **Monitor Account Activity**: Regularly review login attempts and transactions for suspicious activity.

Now that you understand the different types of exchanges and how to stay safe, you have all you need to know to make your first investment. The reality is that, as you get more experience under your belt as an investor, you may use several exchanges. Many investors use a centralized exchange to buy more popular crypto offerings and then utilize DEXs to find more obscure coins and tokens. Choose an exchange, get your account set up, maybe even buy some crypto. In the next chapter, we'll walk through smart investment strategies to help you confidently build your crypto portfolio.

6

CREATING A ROADMAP: INVESTMENT STRATEGIES FOR BEGINNERS

I still remember the beginning of May 2021. Bitcoin hit $63,000 (which was a lot then). Ethereum soared past $3,600 on its way to over $3,900. It was exciting. I was also new to the crypto space and naive.

So I jumped in. All in. With money I couldn't afford to lose. I bought near the absolute peak and then freaked out shortly after.

Bitcoin crashed 50% in days. Ethereum tumbled. The altcoins I'd bought dropped through the floor. My stomach churned each morning as I checked my portfolio. In a matter of a few weeks, I had lost over $10,000. True story.

Why did this happen? Simple. I had no plan. No strategy. No rules. Just raw emotion and the desperate fear of missing out. I wanted financial freedom and someone told me that this was the way to go.

I bought the surge. Then I panic-sold market drops. I had no idea what I was doing. I leveraged positions I didn't understand. Each decision made perfect sense in the moment, driven by either greed or terror. Looking back, I see the pattern of someone gambling, not investing.

This painful experience taught me the most valuable lesson in investing: without a concrete plan, you're just gambling with fancier terminology.

Many beginners make this exact mistake. They react to market movements instead of following a thought-out strategy. They buy because prices rise and sell because prices fall. Emotions drive every decision.

A solid investment plan works differently. It sets clear goals and timeframes. It establishes risk tolerance. It creates rules for entries and exits. It removes emotion from the equation entirely. It creates guardrails that protect you in the most exciting and the most uneasy of times.

If you were going on a road trip, you wouldn't just start driving randomly and hope you'll eventually reach your destination. You'd map the route. Book hotels. Check the weather.

In this chapter, we'll build your investment roadmap. We'll craft rules that work specifically for you: your goals, your timeline, and your comfort with risk. We'll develop strategies that survive both bull and bear markets.

The market will always go up and down. News will always create temporary panic or euphoria. But with a solid plan, these movements become background noise rather than life-altering events.

How to Create Great Investment Goals

Despite my personal horror story, investing in cryptocurrency *can* be an exciting journey. Don't let my story become your story.

Taking some time to reflect on and assess your financial capacity, risk tolerance, and personal objectives are important steps in developing an investment plan that aligns with your long-term wealth-building strategy and short-term financial needs. Let's walk through it together.

Step 1: Determine Your Financial Capacity & Risk Tolerance

Before you put your life savings down on DOGE, take a comprehensive look at your current financial situation. Cryptocurrency markets are volatile. Prices fluctuate dramatically within hours or even minutes. One minute, the market isn't moving at all, and then 30 minutes later, everything has dropped 10%. If you are a fairly risk-averse person, you need an approach that doesn't leave you awake at night sweating at the thought of waking up the next morning to see 30% of your portfolio evaporated. It happens (it also can come back around just as quickly).

How to Assess Your Financial Situation:

1. **Calculate Your Disposable Income:** Determine how much money remains each month after covering essential expenses like rent, utilities, groceries, and savings.
2. **Consider Other Things You Need Money For:** If you have other short-term financial goals outside of crypto, you need to take those into consideration. Do you have at least 3 months of emergency savings? Do you want to diversify outside of crypto? Save for your dream vacation to Europe? Start a family? These are all things you should factor in before you decide how much money you're comfortable investing.
3. **Set an Investment Budget:** Never invest money you can't afford to lose. Start small, gain experience, and gradually increase your investment as your confidence grows.
4. **Understand Your Risk Tolerance:** Ask yourself: How would I react if my portfolio lost 20-50 percent of its value overnight? Can I stomach a bear market without getting scared and dumping your investments? Consider whether you are comfortable with high-risk, high-reward opportunities or if you prefer stable, long-term growth. This will inform how aggressive your investing strategy is.

Step 2: Define Your Crypto Investment Goals

Once you know how much money you're comfortable investing and what types of investment opportunities within crypto are best suited for your risk tolerance, the next step is setting clear, structured goals. The SMART framework is a widely used method for defining effective financial objectives.

SMART Investment Goal Criteria:

- **Specific:** Clearly define what you want to achieve.
- **Measurable:** Ensure you can track progress.
- **Achievable:** Set realistic expectations based on market conditions.
- **Relevant:** Align goals with your financial aspirations.

- **Time-bound:** Establish a deadline for evaluation.

Example of a SMART Crypto Investment Goal:

Instead of saying, *"I want to invest in crypto,"* use the SMART approach:

"I aim to have a crypto portfolio worth $20,000 within the next 2 years."

This goal is specific, measurable, achievable, relevant, and time-bound, making it easier to track and adjust along the way.

I would strongly recommend against creating goals that are centered around achieving specific returns because of how unpredictable financial markets are.

While the example above is results-oriented, having a goal that is centered around a portfolio value gives you a lot of options when it comes to *how* you achieve the goal as long as it's achievable and realistic. Don't set a goal of having a $20,000 portfolio in 2 years if you only have $1,000 to invest. There's a higher chance of you chasing markets and making impulsive decisions that lead to you losing your money than there is of you actually reaching the goal.

I would recommend a process-oriented goal that focuses primarily on actions you can take, such as investing a certain dollar amount each month and achieving a certain portfolio diversification. This allows you to have control over the outcome of the goal and ensures that you stick to a process that aligns with your strategy. It helps keep your emotions out of the mix and has a higher chance of leading to a better outcome over time.

Since all financial markets can be unpredictable and dependent on several different factors that you don't have control over, choosing a results-oriented goal, such as targeting a specific return over a period of time, may not be achievable.

All that being said, investment goals vary based on personal preferences, timelines, and risk tolerance. It's up to you. Just be aware that if you set insane goals, you could get insane results that lead to 200% gains or 90% losses.

Here are some common objectives that investors pursue:

- Long-term wealth building and retirement planning.

Example SMART goal: I want 20% of my entire retirement portfolio to be comprised of cryptocurrency by 2026.

- Generating passive income through crypto.

Example SMART goal: I want a crypto portfolio that produces $1000 each month in passive income by 2027.

- Short-term gains for travel, lifestyle, or side projects.

Example SMART goal: I want a crypto portfolio that produces $5000 in gains each year by the time I'm 40.

Step 3: Break Your Goal into Milestones and Create a Plan to Reach Each One

Once you have a goal in mind that is specific, measurable, achievable, realistic, and time-bound, you can use that goal to reverse engineer your investment plan. You need to have a plan in place that is going to give you a realistic shot at achieving your goal, which means having the right strategy. If you want to achieve a 50% return on your investments in 6 months, you're going to want a different strategy than someone whose goal is to diversify their larger retirement portfolio.

At the end of the day, it's entirely up to you to decide what that strategy is and then periodically evaluate your strategy against your goal and adjust if necessary. So, let's discuss what some of those strategies could be. After reading the following sections you should have a good idea of what type of strategies fit your goal so you can develop your investment plan.

Long-term vs. Short-term Strategies

At the core of any investment plan are two primary approaches:

- **Long-term investing**: Buying and holding crypto assets for an extended period.

- **Short-term trading**: Actively buying and selling to profit from price fluctuations.

Each strategy has its advantages and challenges.

HODLing: Long-term Investing

The term "HODL" originated from a 2013 BitcoinTalk forum post, where a user, in a drunken rant about Bitcoin's volatility, misspelled "hold" as "HODL." What started as a typo quickly became a rallying cry for crypto investors, embodying the philosophy of "Holding On for Dear Life."

HODLing has become a serious long-term investment strategy that has stuck. It encourages investors to ignore short-term market fluctuations and stay committed to their holdings despite market fluctuations.

Understanding the HODL Strategy

At its core, HODLing is about long-term conviction in crypto assets. It works well for investors saving money for medium to long-term goals or retirement. HODLing is not for you if you are investing money you expect to need in the next 5 years.

Key Principles of HODLing

- **Long-Term Perspective**: Focus on multi-year growth, not daily price changes.
- **Emotional Discipline**: Avoid fear-based selling during market dips.
- **Belief in Fundamentals**: The philosophy here is to "buy quality." Invest in cryptocurrencies with strong use cases, technology, and adoption potential.
- **Minimizing Stress**: Reduce the need for constant monitoring and reacting to short-term market fluctuations.

Psychological Benefits of HODLing

Market unpredictability can cause a lot of stress and anxiety. The HODL strategy helps investors stay calm and collected by focusing on long-term goals rather than short-term price movements.

Why Choose Long-term Investing?

- **Reduces FOMO (Fear of Missing Out):** Investors often feel pressure to buy when prices surge. HODLers remain focused on long-term gains rather than chasing what's trending.
- **Avoids Panic Selling:** Emotional reactions to market crashes can result in selling low and missing future rebounds. HODLers trust in their investments and ride out downturns.
- **Takes Advantage of Compounding Growth:** Long-term price appreciation often outweighs short-term fluctuations.
- **Requires Minimal Time Commitment:** Ideal for investors with busy schedules who prefer a passive approach.

Case Studies: Bitcoin & Ethereum HODL Success

Bitcoin (BTC): The Ultimate HODL Asset

- In early 2010, Bitcoin traded for just a few cents, briefly crossing $1 for the first time in February 2011.
- By December 2017, BTC was just under $20,000, and by 2021, it surpassed $60,000.
- In January 2025, BTC hit $109,000.
- Investors who HODLed from the early days have seen exponential returns.

Ethereum (ETH): Patience Pays Off

- Launched in July 2015, ETH initially traded around $0.30–$0.75 on early markets.
- By 2021, ETH reached an all-time high of $4,800.
- As of early 2025, ETH sits at around $2,000.
- Although ETH has clearly dropped significantly from its all-time high, those who believed in Ethereum's smart contract technology and ecosystem growth have benefited significantly since 2015.

How to Implement a HODL Strategy Effectively

Adopting the HODL strategy requires planning and discipline.

Step 1: Set Clear Long-Term Investment Goals:

- Define why you are investing in crypto (e.g., retirement, wealth growth, financial independence). You've done this step already.
- Set realistic expectations. In the short-term, your investments may go down, but long-term adoption trends show consistent growth.

Step 2: Choose Cryptos with Strong Fundamentals

- Research projects with: Real-world use cases (e.g., ADA, LINK, SOL), strong developer activity and adoption, and clear roadmaps and innovative solutions. This is called fundamental analysis and we'll discuss how to do it in chapter 8.

Step 3: Allocate Funds Wisely

- Invest only what you can afford to hold for 5+ years. Consider a diversified portfolio with a mix of blue-chip (Bitcoin, Ethereum) and promising altcoins.

Step 4: Block Out All the Noise

- Don't react to media FUD (Fear, Uncertainty, Doubt) or short-term crashes. Trust your research and stick to your long-term thesis.

Step 5: Revisit Your Strategy Periodically

- Track progress by looking through your portfolio each year. Adjust holdings if fundamentals change. Avoid knee-jerk reactions.

When to Exit a HODL Position

Although HODLing is a long-term strategy, there are times when selling might make sense. Here are some situations where selling is justified:

- **Major Fundamental Changes**: If a project has been around for several years and hasn't gained market adoption or faces regulatory risks.
- **Portfolio Rebalancing**: Adjusting allocations to lock in profits and diversify holdings.
- **Financial Necessities**: If you need funds for real-life expenses, it's okay to cash out strategically. We'll talk tax implications of doing so in chapter 9.

Challenges of Long-term Investing

- **Enduring Market Crashes:** Prices can drop significantly before rebounding. Patience is key.
- **Missed Short-term Gains:** Holding long-term means potentially missing out on quick profit-taking opportunities.
- **Regulatory & Technological Risks:** Crypto is advancing very quickly, and some projects may become obsolete and lose relevance over time as more superior technological innovations hit the market.

Short-term Trading: Capitalizing on Market Unpredictability

Short-term trading involves frequent buying and selling to take advantage of price swings in the market. Unlike long-term investing, short-term strategies require constant monitoring, technical analysis skills, and quick decision-making.

Why Choose Short-term Trading?

- **Potential for Quick Profits:** Traders can capitalize on daily or weekly price fluctuations.
- **Opportunity to Hedge Against Market Downturns:** Profits from short-term trades can offset long-term losses.
- **Involves Active Market Participation:** Appeals to those who enjoy analyzing charts, trends, and news.

Challenges of Short-term Trading

- **High Stress & Emotional Pressure:** Requires constant attention and decision-making.
- **Increased Trading Fees:** Frequent transactions can accumulate significant fees, so be cognizant of the price you are buying and selling at to know what the true profit is.
- **Risk of Major Losses:** Poor timing or emotional trading can lead to substantial financial losses.

Popular Short-term Trading Strategies

1. **Swing Trading:** Buying crypto at a low price and selling it after a few days or weeks when the price rises. Requires technical analysis to identify trends and market cycles.
2. **Day Trading:** Buying and selling crypto within a single day to profit from intraday price movements. Demands fast execution, market awareness, and strict risk management.
3. **Scalping:** Involves making multiple trades within minutes or hours to profit from small price fluctuations. Highly technical and suited for experienced traders.

Long-Term vs. Short-Term: Which Crypto Strategy Fits You?

To summarize, your choice between long-term investing and short-term trading depends on several factors:

1. Investment Goals

- **Long-term:** Building wealth, retirement planning, or passive investing.
- **Short-term:** Generating income, quick profit opportunities, or capitalizing on unpredictability.

2. Time Commitment

- **Long-term:** Minimal time required—check portfolio periodically.
- **Short-term:** Requires constant market analysis and decision-making.

3. Risk Tolerance

- **Long-term:** Suited for risk-averse investors who can handle market dips without panic selling.
- **Short-term:** Appeals to risk-tolerant investors comfortable with quick decisions and price fluctuations.

4. Market Knowledge & Experience

- **Long-term:** Requires basic knowledge of blockchain, crypto fundamentals, and project viability.
- **Short-term:** Demands technical analysis skills, chart reading, and active trading discipline.

Best Practice: If you're unsure, start with long-term investing before exploring short-term trading.

Hybrid Strategy: Combining Long-term & Short-term Investing

Many investors blend both strategies to maximize returns while managing risk. A hybrid approach allows you to benefit from both market cycles and short-term opportunities.

How to Implement a Hybrid Strategy

- 70 percent Long-term Holding (BTC, ETH, Large-cap Coins): Stability and long-term growth.
- 20 percent Swing Trading (Mid-cap Altcoins, Layer 1 & 2 Solutions): Moderate risk and potential gains.
- 10 percent Speculative (NFTs, DeFi, Meme Coins): High-risk, high-reward bets on emerging trends.

Now, look back at the goal you created for yourself. What strategy is going to work best for you? HODL? Trade? Or both?

But choosing a strategy is just the beginning. To give yourself the best chance of success and protect your investments along the way, you'll also want to think about how to build a well-balanced portfolio.

Diversifying Your Cryptocurrency Portfolio

Just as you wouldn't eat only one type of food at a buffet, you shouldn't put all your money into a single cryptocurrency. Even top assets like Bitcoin and Ethereum can swing wildly in price. Diversifying your portfolio helps manage risk and smooth out returns—when one asset dips, others may rise or stay steady. The way you diversify will depend on whether you're a long-term HODLer or a short-term trader.

Why Diversification Matters

A diversified crypto portfolio can:

- **Capture Different Growth Opportunities**: Spreading investments across various types of crypto assets like DeFi tokens, Layer 1 platforms, or NFT-related coins helps you benefit from growth in different areas of the market.
- **Manage Risk Across Market Cycles**: During bull runs, altcoins often outperform Bitcoin (this is called "altcoin season"). But when the market turns bearish, Bitcoin typically holds value better. A mix of assets helps protect you in all conditions.

Instead of betting everything on one project, smart investors build a portfolio with a balance of stability, innovation, and growth potential.

Key Factors to Consider When Diversifying

When selecting cryptocurrencies for your portfolio, it's essential to evaluate assets based on several key factors:

1. **Market Capitalization & Liquidity:** Market cap indicates a cryptocurrency's overall size and stability. Liquidity ensures that assets can be easily bought or sold without significant price impact.
2. **Use Cases & Industry Applications:** Each cryptocurrency serves a different function. Some focus on privacy, others on financial services, gaming, or infrastructure development. Research the

problem each asset aims to solve and whether its technology offers a competitive advantage.

3. **Asset Correlation:** Asset correlation measures how cryptocurrencies move relative to one another. Some coins, like Bitcoin and Ethereum, often move in sync, meaning they won't fully balance risk in a downturn. Ideally, your portfolio should include low-correlation assets that perform independently of each other. If your crypto portfolio is only a portion of your larger portfolio that also includes traditional investments like stocks and bonds, this could be less of a concern for you because crypto doesn't tend to be highly correlated with those markets as of now.

Understanding asset correlation is key to building a balanced crypto portfolio. But how you actually structure that portfolio matters too. Let's look at some practical strategies for integrating crypto into your broader investment plan.

Portfolio Integration Strategies

1. Index Funds and Crypto ETFs

Crypto index products (offered by platforms like Bitwise and Crypto20) aim to track multiple cryptocurrencies, but crypto index funds aren't widely available through traditional brokerage accounts yet.

Bitcoin ETFs—both spot and futures—now offer regulated access to crypto markets. Ethereum futures ETFs are available, but a U.S.-based spot ETH ETF is still pending approval as of early 2025.

2. Core-Satellite Approach

- **Core Holdings**: Traditional investments (stocks, bonds, ETFs) form the portfolio foundation.
- **Satellite Holdings**: Cryptocurrencies provide high-growth potential without compromising stability because they make up a smaller percentage of your portfolio than traditional assets.

3. Dollar-Cost Averaging (DCA) into Crypto

DCA smooths out crypto's price swings by buying on a schedule (like dumping 3% of each paycheck into retirement). You buy when prices soar AND when they crash. No more beating yourself up for buying at all-time highs.

Perfect for investors who know timing markets is basically impossible. And let's face it. That's all of us.

Three Portfolio Styles That Actually Work

1. The Sleep-Well-At-Night Portfolio (Low Risk)

- 60% Bitcoin
- 20% Ethereum
- 20% Stablecoins (USDT, USDC)

Perfect for: People who can't stomach watching their portfolio drop overnight. Long-term believers who just want exposure without the drama.

2. The Growth Machine (Medium Risk)

- 40% Bitcoin
- 30% Ethereum
- 20% Solid Altcoins (ADA, SOL, DOT)
- 10% DeFi tokens (AAVE, UNI, LINK)

Perfect for: Investors comfortable with moderate risk for better returns. Good for medium-term goals or younger retirement savers.

3. The Moonshot Portfolio (High Risk)

- 50% Blue Chips (BTC, ETH, BNB)
- 30% Small-Cap Altcoins
- 20% Speculative Plays - NFTs, dog coins, and gaming tokens that might 100x (or go to zero)

Perfect for: Risk-takers with iron stomachs. The crypto equivalent of skydiving without checking your parachute twice.

BONUS: Using Market Cycles Without Being Psychic

You can't predict markets perfectly, but you can prepare for whatever happens.

HODLers should stockpile during bear markets instead of panic-selling.

Traders can use cycle patterns to time entries/exits better.

Newbies should study market history to avoid emotional mistakes that cost you money.

Managing Your Crypto Like a Pro

Diversification alone won't save you. Add these habits:

- **Rebalance Regularly:** Your 80/20 split today might become 95/5 after a big run-up. Fix it.
- **Stay Informed:** Keep up with news without getting sucked into Twitter drama.
- **Cap Your Risk:** Never put more than 5-10% in moonshots unless you enjoy financial pain.
- **Wallet Security:** Cold storage for your retirement fund, hot wallets for trading money.

Exercise: Create Your Personal Crypto Game Plan

Time to stop reading and start doing.

Step 1: Make a SMART Investment Goal

Choose something Specific, Measurable, Achievable, Relevant, and Time-Bound.

Questions to answer:

- How much will you invest each month?
- Which assets will you focus on?
- What's your timeline for achieving your goal?

Example: *"I'll invest $300 monthly: 50% BTC, 30% ETH, 20% altcoins. Portfolio review every six months."*

Step 2: Pick Your Fighting Style

Match your strategy to your personality and life:

- Do you want to be a passive investor or an active trader? Or are you a twist cone type of person who likes their vanilla with their chocolate?
- How will you handle crashes?

Example: *"I'm going long-term, holding at least three years while staking for passive income."*

Step 3: Build Your Diversification Plan

Ensure your asset allocation supports your goal and risk level.

- Is my portfolio balanced between high-risk and stable assets?
- Do I understand the risks of each investment?

Example: *"To balance growth and stability, I'll allocate 70% to long-term holds (BTC & ETH), 20% to altcoins, and 10% to stablecoins."*

Step 4: Build Your Implementation Plan

Turn your strategy into an actionable timeline. Here's an example for someone with a long-term strategy:

- **0-6 Months:** Learn, track trends, and make small test investments.
- **6-12 Months:** Invest consistently (e.g., $300/month), stake assets, and monitor performance.
- **1-3 Years:** Diversify further, take profits as needed, and rebalance based on market shifts.
- **3+ Years:** Maintain long-term holdings, withdraw profits, and reinvest strategically.

Step 5: Commit to Re-Evaluate Regularly

Set a quarterly or biannual check-in to regularly reevaluate your goal and plan where you:

- Review your risk tolerance and goals.
- Adjust your portfolio as needed.
- Take profits or reinvest based on market conditions.

Not all plans are perfect the first time, and you may find that your goal isn't as realistic as you thought, you're less risk-tolerant than you believed you were, or you don't want to commit the time and energy to learn what's required to be a successful short-term investor. That's okay, just adjust as necessary.

Even with a great plan in place, many beginners fall into common traps that can derail their progress and lead to costly mistakes. In the next chapter, we'll break down the most frequent investing missteps and, more importantly, how to avoid them.

7

COMMON INVESTOR MISTAKES, SCAMS, AND HOW TO AVOID THEM

Floyd Mayweather Jr. stood in the spotlight, flashing his signature smile. The boxing legend held up a Centra Card—a sleek, metallic debit card that promised to let users spend their cryptocurrency anywhere. "I personally own Centra Token," Mayweather boasted on Instagram. "You can call me Floyd 'Crypto' Mayweather from now on."

The year was 2017. A year when crypto money flowed like water.

Behind Centra Tech stood Sam Sharma, Ray Trapani, and Robert Farkas, three twenty-somethings with no background in finance or technology. They claimed their Centra Card could instantly convert cryptocurrencies to cash for purchases at any store. Their ICO (Initial Coin Offering) raised over $25 million from eager investors.

But there was a problem. A big one.

The entire operation was built on lies. The executives listed on their website? Made up. Their partnerships with Visa and Mastercard? Never existed. Their licenses to operate in 38 states? Pure fiction. A working Centra Card? Completely fabricated. The co-founders even created a fake CEO named "Michael Edwards," complete with a stolen photo and made-up biography claiming he had 20+ years of banking experience. The picture of

"Michael Edwards" on the Centra Tech website was actually a college professor living in Canada who had no idea he was the fake CEO of this company.

By April 2018, the FBI arrested Sharma and Farkas. The SEC charged them with fraud. Trapani took a plea deal and received no jail time. Mayweather and DJ Khaled later paid hefty fines for promoting the scam without disclosing they were paid to do so. Investors lost their money. Many never recovered a penny.

If you're interested in learning more about this scam, there is a documentary on Netflix called *Bitconned* that tells the story in much greater detail.

The CentraTech scam highlights something crucial about crypto investing: celebrity endorsements mean nothing. Flashy marketing can mask empty promises. Things that sound too good to be true usually are.

It's easy to think, "I'd never fall for that." But scams like Centra Tech weren't just random internet frauds. They were polished, well-marketed, and fooled even experienced investors. And they weren't alone.

Centra Tech wasn't the only project that took advantage of investors. The ICO boom of 2017 was a gold rush that made some people rich but left many others broke. To understand why, let's look at how ICOs worked and why they became so popular.

What Are ICOs & How Do They Work?

An Initial Coin Offering is a fundraising mechanism used by blockchain-based startups to raise capital before officially launching a new token or blockchain platform without having to rely on banks or venture capitalists. Think of it like Kickstarter for crypto. A team pitches an idea, people invest early, and if the project takes off, those investors make money. But if it fails (or is a scam, which was often the case in 2017), those tokens become worthless.

Between 2017 and 2018, ICOs raised over $20 billion.

This period also exposed major flaws in the system:

1. The Lack of Regulation

- Unlike traditional IPOs (Initial Public Offerings), ICOs lacked government oversight or investor protections.
- This allowed fraudulent projects to launch easily, promising unrealistic returns with little accountability.

2. The Majority of ICOs Failed or Disappeared

- A 2018 report found that almost 80 percent of ICOs launched in 2017 were scams, and another 10 percent failed within a year.
- Many projects never developed functional products, leaving investors with worthless tokens.

Why This Still Matters Today

These scams weren't isolated events. They were part of a pattern that still plays out today in new forms. ICOs may not dominate the crypto space anymore, but the playbook hasn't changed. Scammers still dangle the promise of massive returns, use big marketing pushes, and count on FOMO to lure in investors. If you understand how ICOs fooled people, you'll recognize the same red flags in today's crypto market. Whether it's a meme coin pump-and-dump, a fake DeFi project, or returns that seem too good to be true, the same red flags apply.

Investing safely and intelligently is a theme throughout this book. We've already discussed how to be a smart investor by choosing the right wallets to store your crypto in and protecting your private keys. We will talk about how to prepare for the tax burdens that come with buying and selling crypto so you don't get surprised by an audit from the IRS next year. This chapter is dedicated to other mistakes crypto investors make. Costly ones. The kinds that can wipe out savings, crush dreams, and leave you questioning your judgment. The kinds of mistakes that you need to know. I'm going to show you how to spot these traps before you fall into them.

I believe that most crypto losses don't happen during market crashes. They happen when smart people make avoidable mistakes. They trust the wrong people. They ignore red flags. They rush decisions. They skip basic research. They forget that you only lose your money when you sell the dip instead of buying it.

Common Cryptocurrency Scams & How to Protect Yourself

1. Phishing Attacks

If you've ever even opened an account on a crypto exchange, you've probably received an email that looked like it was from the exchange claiming your account had been locked or compromised, and that you needed to take urgent action to have it reinstated by clicking a link included in the email. Phishing is one of the most prevalent scams in the crypto space for one reason: it works. Scammers send fraudulent emails, texts, or messages that appear to come from a trusted exchange. They can be tricky because a lot of them look legitimate.

These messages often:

- Claim that your account has been compromised or requires urgent action.
- Contain fake login links leading to spoofed websites designed to steal your credentials.
- Request sensitive information, such as your password, private keys, or 2FA codes.

How to Stay Safe:

- Always verify the email sender's address and avoid clicking links in unsolicited messages.
- Manually enter the exchange's URL instead of relying on email links.
- Enable anti-phishing codes (offered by exchanges like Binance) to detect fake emails.

Remember, exchanges will never ask you for sensitive information via email.

2. Fake Exchange Websites

Fraudsters create counterfeit websites that closely resemble legitimate exchanges to trick users into entering their login details. Once credentials are entered, scammers gain access to the real account and steal funds.

How to Identify Fake Websites:

- Look for HTTPS encryption (a padlock icon next to the URL).
- Verify the domain name. Scammers use slight misspellings, such as "binace.com" instead of "binance.com."
- Use bookmarks to save the correct exchange URL for future logins.
- Again, a tool like Wallet Guard can help detect these sites for you.

3. Ponzi Schemes & Fake Investment Platforms

Ponzi schemes lure investors with guaranteed high returns and minimal risk. Early investors are paid using the money from new investors, creating a cycle that collapses once recruitment slows.

Red Flags of Ponzi Schemes:

- Unrealistic returns (e.g., "Earn 10 percent daily with no risk!")
- No clear business model explaining how profits are generated.
- Pressure to recruit others to earn rewards.

How to Stay Safe:

- Be skeptical of any crypto investment that guarantees high returns with no risk.
- Research the team, technology, and business model before investing.
- Check for regulatory compliance and user reviews.

For a textbook example of how Ponzi schemes operate, 'Madoff: The Monster of Wall Street' on Netflix provides an excellent in-depth case study, though it's not from the cryptocurrency space.

4. Pump-and-Dump Schemes

Pump-and-dump schemes occur when fraudsters artificially inflate an altcoin's price through misleading publicity and fake endorsements. Once prices surge, they sell their holdings, causing prices to crash and leaving late investors with losses.

How to Stay Safe:

- Avoid investing based on social media trends.
- Research market fundamentals before buying altcoins.
- Be wary of new tokens with sudden, unexplained price spikes.

While understanding common scams is important, protecting yourself also comes down to the tools and platforms you use. You can keep yourself safe by implementing the exchange security best practices we discussed in chapter 5. The other best defense against scams is education and staying informed about what's happening in the crypto world.

That means learning how to track trends, follow credible news sources, and spot red flags before they become costly mistakes.

Tracking Crypto Trends and News

Where does a person get accurate information from? There are lots of places. You just have to know where to look.

Finding Trustworthy Information

Information quality varies widely in crypto. Some sources provide accurate reporting. Others spread misinformation for clicks or manipulation. Here's where you can find information you can count on.

1. Crypto-Focused News Sites

CoinDesk, Decrypt, and CoinTelegraph have established themselves as leading crypto news platforms. CoinDesk is respected for its market analysis and regulatory coverage, often breaking major industry stories. CoinTelegraph provides thorough articles on blockchain technology and emerging trends. Decrypt is great for both new and experienced investors because its content strikes a balance between education and breaking news.

These sites provide deeper insights into blockchain developments and market movements than general news outlets because they focus solely on cryptocurrency.

2. Traditional Financial Media

Traditional finance outlets have increasingly dedicated resources to crypto coverage:

- **Bloomberg's Crypto Section** views digital assets through a global economic lens, placing developments in broader market context.
- **CNBC Crypto** features interviews with industry experts and institutional analysis of major trends.
- **Forbes Crypto** focuses on institutional adoption and regulatory frameworks shaping the industry.

These mainstream outlets connect crypto trends with wider economic events. This is great in helping investors understand correlations with traditional markets.

3. Real-Time Tracking Tools

Since crypto markets never close, real-time monitoring tools help investors stay informed around the clock.

Several mobile applications give investors instant market access:

- **Blockfolio (FTX App)** tracks portfolio performance and broader market trends.
- **CryptoCompare** offers customizable price alerts and detailed asset comparisons.
- **TradingView** provides advanced charting tools and technical indicator-based alerts for more sophisticated analysis.

For staying on top of specific developments, Google Alerts lets you set notifications for particular topics like "Bitcoin regulation" or project names. Feedly aggregates news from multiple sources into a single dashboard, creating a personalized news feed filtered to your interests.

4. Social Media & Community Sources

Great information can also be found on social media and in the wider crypto community. But these places are also where you need to have your guard up because they're also breeding grounds for bad information from rookie investors posing as experts, scammers, and other bad actors looking to manipulate markets for their own gain.

- **X (formerly Twitter)** acts as the crypto market's pulse. Here, you can find real-time updates from influencers, analysts, and developers. Breaking news about regulations, hacks, or partnerships often appears on X first, and discussions there can directly influence market sentiment. The platform's immediacy makes it valuable but requires careful filtering. You need to be aware of hype and paid promotions disguised as expert advice.
- **Reddit** communities, particularly subreddits like r/cryptocurrency, offer in-depth discussions where users share technical analysis and personal investment experiences. These forums enable debates and different viewpoints, creating a more balanced perspective than individual accounts can provide. If a forum is like an echo chamber where skepticism is encouraged, that's a red flag.

Pro Tip: Monitor engagement levels and emotional tone in communities you follow. These social signals can provide early warnings about potential market turning points.

- **Telegram and Discord** host private communities with exclusive discussions and direct access to project teams. Many crypto projects maintain official channels where developers answer questions directly. Some trading-focused groups provide alerts about market movements and whale activity tracking. Be aware, these are also breeding grounds for pump-and-dump schemes. Always verify information before acting.

Key People to Follow

Following the right voices can give you early insights into trends before they hit mainstream media. Crypto founders often share technical updates, while analysts break down price movements in real-time. Here are some notable names you can rely on for good information:

- **Vitalik Buterin:** Regularly discusses Ethereum's roadmap, scaling solutions, and broader crypto philosophy.
- **Changpeng Zhao (Former Binance CEO):** Shares perspectives on exchange trends and what's going on across the industry.
- **Brian Armstrong (Coinbase CEO):** Offers views on institutional adoption, regulatory challenges, and exchange operations

Beyond founders, market analysts like **CryptoCred** provide technical analysis and market education that helps traders develop their strategies. On-chain analyst **Willy Woo** specializes in Bitcoin price trends using blockchain metrics rather than traditional technical analysis. **Raoul Pal** bridges traditional finance and crypto by covering macro-financial trends affecting digital assets, drawing on his background in global macro investing.

5. ChatGPT Prompts to Keep You Updated

If you're at all familiar with ChatGPT or other chatbots like Claude, these tools can help you aggregate market news in seconds rather than hours. You can ask a chatbot for assistance with things like:

Daily or Weekly Market Summaries

Stay informed about the latest price movements, news, and trends.

- **Prompt:** *"Give me a summary of today's top crypto news."*
- **Prompt:** *"Summarize key crypto market trends over the past month."*

Tracking Fear & Greed Index Movements

Understand market sentiment shifts and how they affect price action.

- **Prompt:** *"How has the Fear and Greed Index changed over the last 6 months? What trends can we see?"*

Monitoring Regulatory and Industry News

Stay ahead of government policies, exchange rules, and major legal changes.

- **Prompt:** *"Summarize recent U.S. crypto regulations and how they impact investors."*
- **Prompt:** *"How are governments reacting to Bitcoin ETFs and institutional adoption?"*

Analyzing Market Sentiment

Gauge investor emotions through social media and news sources.

- **Prompt:** *"What is the general sentiment around Ethereum on social media this week?"*
- **Prompt:** *"Summarize Bitcoin discussions on Reddit and Twitter in the last 7 days."*
- **Prompt:** *"Analyze the latest sentiment on Solana and whether it's bullish or bearish."*

Pro Tip: Ask ChatGPT for sources so you can read articles on the topics you inquire about. ChatGPT has also known to provide inaccurate information, so being able to go straight to the source can be really helpful.

Following market news and developments helps you spot trends before they fully develop. Reliable news sources and market analysis are key to this skill because crypto thrives on online community involvement in ways traditional markets don't. These communities provide immediate access to news, analysis, and market sentiment shifts that often precede price movements.

How Crypto News Impacts Prices

Institutional adoption drives significant price movements in crypto. When major companies like Tesla purchased Bitcoin in 2021, the price jumped dramatically as investors gained confidence. Bitcoin surged when El Salvador made it legal tender. Similarly, markets rise when governments create crypto-friendly regulations because they provide clarity and security. People know what's okay and what's not okay, and since the government is recognizing crypto as needing regulation, it legitimizes the assets. Technological improvements also boost investor interest; Ethereum's switch to Proof-of-Stake generated excitement about the network's future and sustainability.

On the flip side, regulatory crackdowns can cause panic selling across markets. China's 2021 Bitcoin mining ban sent shockwaves through the entire ecosystem, forcing miners to relocate and creating temporary market uncertainty. Security breaches at exchanges or protocol vulnerabilities significantly damage investor trust in institutions and often lead to immediate sell-offs. Broader economic factors like inflation, interest rate changes, or financial instability tend to increase crypto price instability as investors reassess risk tolerance.

Example: The FTX collapse in November 2022 triggered a broader market downturn, contributing to an estimated $150–200 billion drop in overall crypto market cap within days.

Pro Tip: Track how markets respond to different news types to better understand price patterns. This awareness helps you anticipate market reactions rather than simply responding to them.

Following crypto news and understanding its market impact is essential. But knowing *how* you react to that news is just as important. In a fast-moving, emotionally charged market like crypto, even well-informed investors can make costly mistakes if they let fear or excitement take over.

Market-Related Mistakes

Psychology plays a huge role in investment decision-making. Investors often repeat the same psychological mistakes throughout multiple cycles of volatility. Learning to recognize these emotional patterns so you can manage your emotions, maintain discipline, and stick to a strategy is critical to your long-term success in the crypto market.

1. Fear & Panic Selling During Market Crashes: When Bitcoin or altcoins drop 20-30 percent in a day, people get scared. Investors panic and sell at a loss because all they're thinking about is not losing the rest of their money. They're not thinking about Newton's Law and the fact that what goes up must come down (mostly) and vice versa. Market dips are often temporary. Those who sell emotionally miss out on long-term gains.

Lesson: Selling based on fear locks in losses. Patience and perspective are key.

2. FOMO & Buying at Market Peaks: People get excited when prices skyrocket. Too excited sometimes. Investors rush to buy, afraid of missing potential profits. What often happens is that people watch prices go up and forget that, at some point, those prices will hit a ceiling (called resistance). Often, this leads to buying at inflated prices before a correction.

Lesson: Don't chase trends.

3. Overtrading Due to Market Noise: Constantly buying and selling based on short-term movements leads to high fees and losses.

Lesson: Educate yourself on market cycles so you can stay cool-headed during dips and crashes.

4. Following Rumors Without Research: Many investors buy altcoins based on X/Reddit trends without due diligence. This leads to losses when the trends fade and the project lacks real utility.

Lesson: Always DYOR. Always.

Other Things to Do Instead

Here are some techniques to manage emotions & maintain investment discipline:

1. Set Predefined Entry & Exit Points: Decide beforehand at what price you will buy or sell to remove emotional decision-making from the equation.

2. Shift to a Long-Term Investment Mindset: Develop habits that keep emotions in check. I could tell you to meditate and take deep breaths, but that hasn't helped me in the past. The thing that has helped me the most is focusing on long-term trends. Review historical data to understand that fluctuations are normal. Strong projects tend to recover and grow over time. Don't obsess over daily price movements, and don't even look at your investments during periods of volatility if you're holding long-term.

3. Don't Put Yourself in Stressful Situations: If you can't stomach deep drops in the market, short-term trading may not be the investing strategy for you.

4. Don't Invest Money You Can't Afford to Lose: Period.

If you are panicking over your investments, this is a good sign that you need to re-evaluate. Did you invest money you need for other things? Are you using the wrong investment strategy for your risk tolerance? Are you just inexperienced, and you jumped into the deep end too quickly?

Answering these questions for yourself can help you figure out how to make a course correction.

Pattern recognition is yet another way to steel yourself against scary markets.

Identifying Market Patterns

We know from history that financial market cycles can act a lot like natural tides going in and out. Each phase has its own rhythm that repeats over time. Even though crypto history is short, this concept still applies. These market cycles generally follow a predictable pattern, composed of four key phases:

- **Accumulation:** Smart investors buy quietly when prices are low.
- **Expansion:** Optimism grows, attracting more buyers and pushing prices higher.
- **Distribution:** Hype peaks, early investors cash out, and markets become overvalued.
- **Decline:** Reality sets in, panic selling kicks in, and prices drop. The cycle resets.

Recognizing these stages is a practical way to approach investing strategically. When you know what phase in the cycle the market is currently in, you can more easily act according to common investing principles: buy low, sell high.

Analyzing past cycles can provide insight into when a market is overheating or undervalued. While no two cycles are identical, they often follow similar psychological patterns. Tracking previous peaks and corrections can help in determining optimal entry and exit points.

Using the Fear and Greed Index as a Tool

Investor sentiment plays a huge role in market fluctuations. The Fear and Greed Index is an easy way to get a good pulse on current market psychology:

- **Extreme Greed:** Signals potential overvaluation. Consider caution.
- **Extreme Fear:** Indicates market undervaluation. Often a buying opportunity.

Buying during fear and selling during greed has historically been a winning strategy. A study found that using the Bitcoin Fear and Greed Index as a trading signal significantly outperformed a buy-and-hold strategy. By investing when the index showed extreme fear and selling during extreme greed, the strategy achieved a 1,145% gain, compared to 235% for buy-and-hold over the same period.

The single most important takeaway from this chapter is simple: thorough research and experience protect you from bad investments. Taking time to check facts, verify claims, and seek multiple perspectives creates a shield against the common pitfalls in crypto markets.

Avoiding scams protects your money, but to grow your wealth, you need to recognize which projects are actually worth investing in. Next, we'll break down how to spot crypto opportunities with strong fundamentals so you can make smarter, more confident investments.

In the next chapter, we'll explore these fundamental analysis skills that form the backbone of smart crypto investing. You'll learn practical techniques to assess a project's real value beyond the market noise.

8

FUNDAMENTAL ANALYSIS: HOW TO SPOT EMERGING CRYPTOCURRENCIES WITH HUGE GROWTH POTENTIAL

IF YOU WERE LATE TO THE BITCOIN GAME, THE REALITY IS YOU'RE NOT GETTING THE same gains as the early adopters who bought for $0.10, $1.00 or even $1,00o per BTC. But does that mean you missed the boat? Hardly.

While Bitcoin created digital scarcity and Ethereum brought programmable money, as we discussed in our conversation on altcoins, newer projects are tackling specific problems with targeted solutions. Some focus on privacy, others on scaling, cross-chain interoperability, or real-world asset tokenization.

Solana rose from around $2 in early 2021 to an all-time high of nearly $260 later that year. Avalanche jumped from under $4 in early 2021 to around $135. Polygon climbed from about $0.01 in 2020 to over $2.90 at its peak. Early investors were rewarded with huge returns because they recognized fundamental value before the market caught up.

But for every success story, fifty projects fade into obscurity. Your job isn't to invest in everything—it's to develop a filter that catches diamonds while screening out the endless stream of garbage. It's about developing the skills to spot solid projects with staying power before they become obvious to the masses. By the end of this chapter, you'll know exactly how to separate the next big thing from the next big nothing.

Let's break down what makes coins valuable. Ready to stop gambling and start investing? Let's get started.

Fundamental Analysis Basics

When you buy crypto, you're not just buying lines on a chart. You're buying into a vision, a technology, and a solution. While those green and red candles might hypnotize day traders, smart money looks deeper.

Think of fundamental analysis (FA) as detective work. You're gathering clues about what a crypto project is really worth. Beyond the price tag it's wearing today. Unlike stocks, where you can obsess over P/E ratios and quarterly earnings, crypto requires a different approach.

You need to examine the code, not the corporation. The community, not the CEO. The problem it solves, not the profit it reports.

Why bother with all this research? Because prices lie. They can fluctuate based on tweets, Trump, FOMO, and market manipulation. But fundamentals tell the truth about a project's actual value. When you spot that disconnect between price and value, that's where money is made.

Let me show you exactly what to look for when sizing up a crypto project.

Key Components of Fundamental Analysis in Crypto

Think of fundamental analysis as a three-legged stool. If any leg is weak, the whole thing can collapse. Here are the three pillars that hold up any solid crypto project:

1. **Technology & Infrastructure**: Is the blockchain secure? Can it handle growth? Does it bring something new to the table?
2. **Utility & Adoption**: Does it solve real problems for real people? Are folks actually using it?
3. **Financial Metrics & Market Position**: Do the numbers make sense? How does it stack up against competitors?

Let's break down each of these pieces so you can spot winners before the crowd catches on.

1. Technology & Infrastructure: The Foundation of a Cryptocurrency

The tech stack is where the rubber meets the road. Great marketing can't save bad code. When examining a project's technology, ask yourself:

- **Is the security legit?** How has it held up against attacks? Has it been audited by reputable firms?
- **Can it scale?** Many projects work fine with a few thousand users but crumble under real adoption. Check transaction speeds, costs, and how it handles network congestion.
- **What's the secret sauce?** If a project is just copying Bitcoin or Ethereum with a few tweaks, that's a red flag. Look for genuine innovation that solves existing problems.
- **How active is development?** Check GitHub commits and updates. A project gathering dust isn't worth your dollars.

2. Utility & Adoption: Real-World Use Cases & Demand

A beautiful blockchain that nobody uses is like a Ferrari gathering dust in a garage: impressive but pointless. The best crypto projects solve actual problems that people care about.

Ask yourself these questions:

- **What itch does this scratch?** If you can't explain the problem it solves in one sentence, there might not be one.
- **Who's using it right now?** Not promises of future users. Actual users today. Check transaction counts, daily active addresses, and app downloads.
- **Are real businesses building on it?** Partnerships with established companies aren't just for show. They signal that professionals bet their business on this tech.
- **How sticky is it?** If users can easily jump to a competitor, they probably will. Look for projects that create lock-in through unique benefits.

3. Financial Metrics & Market Position: Evaluating Growth Potential

Numbers don't lie. While crypto isn't all about financials, they matter. Here's what to check:

Market Cap = Coin Price × Circulating Supply

That $0.00001 token isn't necessarily cheaper than Bitcoin. Market cap tells the real story. A $50 million market cap coin has much more growth potential (and risk) than a $50 billion one.

Token Supply Mechanics

- **Fixed Supply (like Bitcoin's 21M cap):** Creates scarcity, potentially driving price up over time.
- **Inflationary Model:** New tokens constantly created can dilute your investment unless demand keeps pace.

Ask: "Are new tokens flooding the market faster than demand can absorb them?"

Liquidity & Trading Volume

Low daily trading volume is a red flag. It means you might not be able to sell when you need to (at least not without tanking the price).

For example, if you spot a coin with decent tech but realize its daily trading volume is barely $50,000, that's too thin. When market sentiment shifts, holders likely wouldn't be able to exit without 40% slippage. Not a position you want to be in.

Using Fundamental Analysis to Identify Strong Investments

Let's get practical. Here's your step-by-step guide to researching a crypto project. This works for established and emerging projects:

1. Read the Whitepaper

The whitepaper is the project's blueprint. It should clearly explain:

- The specific problem being solved (not vague phrases like "revolutionizing finance")

- How the solution works—technically and practically
- A realistic roadmap with defined milestones
- Token utility and role in the ecosystem

Red flags: Excessive jargon, unrealistic promises, missing technical detail, or tokenomics that feel bolted on just to create hype.

2. Research the Team & Development Activity

A crypto project is only as strong as the people building it.

- Look up team members on LinkedIn and Google. Have they built credible things before?
- Check GitHub activity (this is where code gets committed). Are they shipping code regularly or just saying they are?
- Watch how founders communicate. Are they transparent about challenges, or just hyping token price?

Bonus tip: No named team? That's a major red flag unless it's a truly decentralized or community-driven protocol with a long track record.

3. Evaluate Product-Market Fit & Competitiveness

Crypto doesn't exist in a vacuum. Ask yourself:

- What makes this project stand out from its competitors?
- Does it offer something new or solve a known problem better?
- Is it a first mover, or chasing others already gaining traction?
- Are there regulatory concerns specific to this niche or jurisdiction?

Remember: the best tech doesn't always win—but the best combination of product, community, timing, and adaptability often does.

4. Assess the Community & Partnerships

Strong communities are a project's lifeline during bear markets. Look for:

- Active discussion around utility, tech, and governance (not just price talk).

- Organic engagement, not inflated numbers from bots or giveaways.
- Real partnerships: verify claims by checking if the "partner" acknowledges them.

Communities focused on long-term goals tend to stick around. Those built on hype tend to vanish.

5. Analyze Tokenomics & Market Potential

Token design can make or break a project. Key questions to ask:

- How is the token distributed? Heavy allocations to insiders or VCs with short lockups can be risky.
- When do tokens unlock? Sudden unlocks can flood the market and crash prices.
- What's the current market cap, and how big is the addressable market? A $50M project targeting a $10B industry may have room to grow. A $5B meme coin with no use case... maybe not.

Pro Tip: Treat evaluating a crypto project like conducting due diligence on a startup. If you wouldn't back it as a business with real money, don't buy the token just because it's trending.

While fundamental analysis gives you the big picture of a project's value, on-chain analytics lets you verify that value in real-time by tracking what is actually happening on the blockchain. It's like moving from reading the business plan to watching the cash register.

On-Chain Analytics: Blockchain Data Insights

On-chain analytics examines actual transaction data recorded on the blockchain. This gives cryptocurrency investors unique advantages not available in traditional markets.

Key On-Chain Metrics Worth Tracking

The most useful blockchain data falls into three categories:

1. Transaction Activity

Transaction Volume: This shows the total value transferred on a network daily. Higher transaction volume typically indicates active usage and real economic activity.

Active Addresses: This counts unique wallet addresses interacting with the network. Growing active addresses often signal increasing adoption and user engagement.

Looking at both metrics together provides better insights than price alone. For example, rising transaction volumes alongside stable prices might indicate building momentum before a price increase.

2. Network Security Data

Hash Rate (For Bitcoin and other Proof-of-Work chains): Hash rate measures the total computing power securing the network. Higher hash rates indicate stronger security and miner confidence.

Staking Percentages (For Proof-of-Stake networks): For networks like Cardano or Ethereum 2.0, this shows what percentage of tokens are locked for network validation. Higher staking rates often indicate investor confidence.

3. Investor Behavior Signals

Exchange Inflows and Outflows: Tracking how much cryptocurrency moves to and from exchanges provides clues about potential selling or buying pressure:

- Large inflows to exchanges often precede selling.
- Significant outflows from exchanges typically indicate long-term holding intentions.

Whale Transactions: These track movements from addresses holding large amounts of cryptocurrency. Sudden increases in whale activity can precede major market moves.

Practical Applications of On-Chain Data

Identifying Bullish Conditions

- Several on-chain signals often appear before price increases: Coins moving from exchanges to private wallets.
- Increasing active addresses despite flat prices.
- Rising transaction values showing growing usage.
- Long-dormant addresses becoming active again (often to accumulate more).

Spotting Potential Bearish Trends

Warning signs worth monitoring include:

- Large amounts of cryptocurrency moving to exchanges.
- Declining network usage despite rising prices.
- Miners or validators selling their rewards at increasing rates.
- Decreasing new address creation.

Tools for Accessing On-Chain Data

Several platforms make this data accessible without technical expertise:

- **Glassnode:** Provides comprehensive metrics with free basic data.
- **Santiment:** Offers unique insights combining on-chain and social media data.
- **CryptoQuant:** Specializes in exchange flows and miner behavior.
- **IntoTheBlock:** Uses machine learning to generate signals from on-chain data.

Even basic weekly checks of these metrics can provide useful insights for investment decisions.

This concludes our exploration of analytical methods for cryptocurrency investing. By combining fundamental analysis, technical chart reading, and on-chain data tracking, you can develop a comprehensive approach to evaluating crypto assets beyond just following market hype.

Understanding what makes a project valuable is only part of being a smart investor. The other part is knowing what the government expects when you make those gains. Before you celebrate your next 5X, let's talk about crypto taxes: what counts as income, what triggers a taxable event, and how to stay compliant without losing your mind.

9

CRYPTO TAX BASICS: HOW TO FOLLOW THE RULES AND MAYBE SAVE SOME MONEY TOO

CRYPTOCURRENCY TAX ENFORCEMENT HAS BECOME A TOP PRIORITY FOR THE IRS IN recent years. What started as a simple checkbox on tax forms has evolved into strict reporting requirements, data-sharing agreements with exchanges, and increased audits.

Starting in 2026 (for the 2025 tax year), your crypto trades will be reported to the IRS whether you were the one who reported them or not. Many investors face penalties not because they're trying to avoid taxes but because the rules are complex and constantly changing.

Every crypto transaction—buying, selling, swapping, or earning—has tax implications. But if you understand the rules, there are also opportunities to reduce your tax burden.

Regulations vary by country, and keeping accurate records is necessary to avoid future headaches. In this chapter, we'll break down the key tax considerations so you can confidently manage your investments without getting caught off guard.

Which Crypto Activities Trigger Taxes?

Not everything you do with crypto creates a tax bill. Knowing what does and doesn't require reporting helps you plan better and avoid surprises.

Actions That Create Tax Bills

When you sell Bitcoin for dollars, it's a taxable event. The same applies when you swap ETH for SOL, buy a new laptop with crypto, or receive mining rewards. Each creates either capital gains (if you sell for more than the purchase price) or ordinary income (from earning new coins).

Even those "free" airdrops come with a price tag. Their market value counts as income the day they land in your wallet. Likewise, payments for services in crypto count as income at their fair market value when received.

What You Can Do Tax-Free

Simply buying crypto with cash won't trigger taxes. Neither will moving coins between your own wallets. HODLing through market swings stays tax-free until you sell, trade, or spend. Wallet-to-wallet transfers might look suspicious to tax authorities initially, but good records prove they're non-taxable.

This distinction matters because tracking only taxable events simplifies your record-keeping burden significantly. The catch? You need to accurately identify which is which.

Given how easily crypto transactions accumulate, keeping track of taxable and non-taxable events can become overwhelming. That's where automation comes in. Instead of manually sorting through every transaction, specialized crypto tax software can do the heavy lifting for you so you can just focus on trading.

Automating Your Crypto Tax Reporting

Manual tracking is a nightmare if you've made more than a few trades. Think about it: each swap between coins, each purchase, each staking reward. They all need documentation. Even small traders can rack up hundreds of taxable events yearly. Tax software helps sleep come easier at tax time.

Popular tools like **Koinly, CoinTracker, ZenLedger, and CryptoTrader.Tax** connect directly to exchanges like Binance and Coinbase. They pull your history automatically, calculate your gains and losses, and spit out ready-to-file forms. Some even track those tricky DeFi transactions and NFT trades that standard tax software misses completely.

The time you save is worth the subscription cost. More importantly, these tools catch transactions you might forget, potentially saving you thousands in audit penalties later.

How Different Countries Tax Your Crypto

No two countries treat crypto exactly the same way for taxes. Some see it as property, others as a financial asset, and the classification directly impacts what you'll pay. Here's a snapshot of tax regulations by country:

United States

The IRS views your crypto as property like real estate or stocks, not as actual currency. This means every sale, trade, or purchase using crypto potentially triggers capital gains tax. Buy a $7 coffee with Bitcoin that's worth more than when you got it? That's a taxable event.

Mining rewards, staking income, and airdrops all count as ordinary income the moment you receive them, even if you never sell. Their value when you receive them becomes your cost basis for later capital gains calculations.

For gifts, you can give up to $19,000 (2025 threshold) in crypto per person yearly without tax consequences. Larger gifts require filing forms.

United Kingdom

HMRC takes a similar but distinct approach. Capital gains apply when you sell or trade, while income tax hits mining and staking rewards. UK tax authorities expect detailed records—dates, amounts, and who was involved in each transaction.

Canada & Australia

In Canada, only half your capital gains face taxation, but the CRA watches closely for patterns that suggest you're trading as a business rather than an investor.

Australia offers a small break. Personal purchases under AUD 10,000 might avoid tax entirely, but the ATO still counts staking and mining as taxable income right away.

Capital Gains: The Core of Crypto Taxes

When you sell or trade crypto at a profit, capital gains tax kicks in. How much you'll pay depends largely on how long you held the coins before selling.

Short-Term vs. Long-Term Holding

The tax difference between selling after 11 months versus 13 months can be dramatic. In the U.S., coins held less than a year are taxed as ordinary income. If you're the 37% tax bracket, you pay 37% on short-term gains. Hold longer than a year, and rates drop to 0%, 15%, or 20%, depending on your income bracket.

This creates a clear strategy: when possible, wait for the one-year mark before selling profitable positions. The tax savings can sometimes outweigh potential market losses from waiting.

Calculating Your Gain or Loss

The math starts with your "cost basis," or what you paid originally, including fees. Subtract this from your selling price to find your taxable gain or deductible loss.

But which coins did you sell? If you bought Bitcoin at $10,000, $30,000, and $40,000, then sold some at $50,000, your tax bill changes dramatically based on which purchase you count against the sale:

- **FIFO (First-In, First-Out):** The IRS default. Your oldest coins ($10,000) sell first, creating a $40,000 gain.
- **LIFO (Last-In, First-Out):** Your newest coins ($40,000) sell first, creating a $10,000 gain.

107

- **HIFO (Highest-In, First-Out):** Your most expensive coins ($40,000) sell first, minimizing your taxable gain.

The method you choose can mean thousands in tax differences. Most countries require consistency, so pick carefully and stick with it.

What Happens If You Don't Report Crypto Taxes?

Tax authorities worldwide have sharpened their focus on crypto. The days of operating under the radar are over. The IRS has demanded transaction data from major exchanges like Coinbase and Kraken. They're matching this information against tax returns to flag discrepancies. Advanced tracking tools give tax agencies the ability to trace transactions across wallets and exchanges.

Some investors mistakenly believe their self-custody wallets remain invisible to tax authorities. In reality, the on-ramps and off-ramps where crypto connects to traditional banking create traceable patterns. When you eventually convert back to dollars or make large purchases, those disconnects become visible.

The Cost of Getting Caught

The penalties hurt. If you underreport crypto income, you may face an accuracy-related penalty of 20%, or up to 75% in cases of proven tax fraud. Accidental underreporting might cost you 20-40% of the unreported amount.

Intentional evasion crosses into criminal territory. One investor faced a three-year prison sentence after hiding over $20 million in crypto gains. Even smaller cases can result in garnished wages, liens on property, and frozen accounts.

The simplest solution? Disclose everything, even if you can't pay immediately. The IRS offers payment plans and voluntary disclosure programs, but failing to report at all can result in penalties, audits, and even criminal charges.

The penalties for non-compliance make a strong case for proper reporting. But how exactly do you prepare everything the IRS wants to see? Getting organized before tax season is the key to both compliance and peace of mind.

Getting Your Crypto Tax Documents Ready

Proper records make all the difference between a smooth tax filing and a stressful scramble. Tax authorities want to see your complete transaction history, not just summaries.

What You Need to Gather

Start by downloading transaction histories from every exchange you've used. Coinbase, Binance, and others offer tax reports but check them carefully. They often miss transfers between your own wallets or have other gaps.

Keep records of any crypto you received as payment for work, mining rewards, or airdrops. For NFTs, document both acquisition and sale dates, along with all associated fees.

The trickiest part? Establishing your cost basis for each transaction. Without this starting point, you can't accurately calculate gains or losses.

Some exchanges issue Form 1099 forms (like 1099-MISC or 1099-B), but reporting practices vary widely and may not reflect full gain/loss information. Don't assume these forms contain everything the IRS needs.

The Record-Keeping Time Trap

The biggest mistake new crypto investors make is waiting until tax season to organize their records. Start tracking from day one. Set up a simple spreadsheet or use dedicated software to log each transaction as it happens. Ten minutes weekly saves hours of frustration (and possibly thousands in overlooked deductions) later.

If you've already fallen behind, block a full weekend to catch up before approaching your tax preparer. They'll charge substantial premiums to sort through disorganized records. If they'll take your case at all.

Professional Help Pays for Itself

If all of this seems overwhelming to you, the best thing you can do for yourself is find a tax professional who understands both crypto and tax law. This often saves more than their fee.

When searching for help, ask about their experience with DeFi protocols, staking taxation, and NFT transactions. A good crypto tax professional stays current on emerging regulations and can help you develop strategies aligned with your investment goals.

More Strategies On The Way

While understanding crypto tax basics is essential, knowing how to minimize your tax burden is equally important. In the next chapter, we'll explore specialized strategies, including tax-loss harvesting and using tax-advantaged accounts like self-directed IRAs. These approaches can potentially save you thousands of dollars each year on your crypto investments.

10

TAX-FREE CRYPTO INVESTING AND LONG-TERM PLANNING

NOW THAT YOU UNDERSTAND HOW TAXES IMPACT YOUR CRYPTO INVESTMENTS, LET'S focus on strategies to legally minimize them. Tax-advantaged accounts offer a powerful solution: they're not just for traditional investments anymore. By strategically using Individual Retirement Accounts (IRAs), Roth IRAs, and Health Savings Accounts (HSAs), you can build your crypto portfolio while significantly reducing or even eliminating tax burdens. Think of these accounts as shields that protect your digital asset gains. This chapter explores practical ways to integrate crypto into these tax shelters and maximize your long-term growth potential.

Crypto IRAs: Merging Retirement Planning with Digital Assets

A Crypto IRA combines retirement savings with digital asset investing. Unlike regular crypto trading, these accounts let you invest in Bitcoin, Ethereum, and other cryptocurrencies without the usual tax headaches.

Traditional IRAs offer tax-deductible contributions with growth that's tax-deferred until withdrawal. This means you'll pay taxes later when you take distributions in retirement.

Roth IRAs work differently. You contribute after-tax dollars and all future growth and qualified withdrawals are completely tax-free.

The Million-Dollar Tax Hack: A Hypothetical Scenario

Consider if you took your $7,000 Roth IRA contribution and went all-in on Solana at $150.

If Solana hits even a fraction of what Bitcoin has done and reaches $25,000 per coin by 2055, that $7,000 investment turns into over a million bucks. The kicker? Because it's in a Roth IRA, you won't pay a single penny in taxes when you withdraw it.

The same investment in a traditional IRA would get hammered with taxes. At a modest 24% tax bracket in retirement, that's $280,000 gone to Uncle Sam.

Obviously nobody knows where crypto prices are heading next month, let alone in 30 years. Some people think Solana will be at $5,000, others think it will hit $70,000. No matter where it lands, it's a win and this much is clear: paying taxes on $7,000 now beats paying taxes on a million dollars later.

Pro Tip: If you're investing for retirement and you expect your tax bracket to be 24% or higher, you would be better off not putting your crypto investments into a traditional IRA. Why? Those gains get taxed as ordinary income when you withdraw them. If your tax bracket is going to be higher than the 20% long-term capital gains tax, your tax bill is going to be higher. As of right now, anyway. Who knows what the tax laws will be next year, let alone 2055?

Setting up a crypto IRA requires finding a specialized custodian, as not all financial institutions support digital assets. Once you select a provider, you can fund your account with cash or roll over money from an existing retirement account. You'll then purchase crypto assets within the IRA structure, maintaining all the tax benefits of the account.

If you want a Roth IRA, just make sure the provider you choose offers them. It's also good to do your due diligence and look to see that the platform offers the cryptos you want to invest in.

For 2025, you can contribute up to $7,000 per year to an IRA, or $8,000 if you're 50 or older. One important limitation: you can't transfer existing crypto directly into an IRA. You must buy new crypto inside the account with contributed cash or cash from a rollover. Also, make sure your custodian offers strong security measures to protect your digital holdings.

Health Savings Accounts (HSAs): A Hidden Gem for Crypto Investors

Most people think of HSAs as boring medical expense accounts. They're missing out on one of the most powerful tax tools in existence. Health Savings Accounts offer something no other investment account does: a triple tax advantage.

What makes HSAs so special? First, your contributions are tax-deductible, lowering your taxable income right away. Second, everything inside the account grows tax-free. Third, when you use the money for qualified medical expenses, you pay zero taxes on withdrawals. That's right. You *never* pay taxes on that money at any point!

The HSA Secret Most People Miss

Consider this real example: you contribute $4,300 to your HSA this year and invest it in Ethereum. Over the next decade, imagine that investment grows to $43,000. Not only did you get a tax break on your initial contribution, but all $38,700 in gains remain completely tax-free when used for healthcare costs.

The best part? Medical expenses are inevitable for everyone. Unlike other tax-advantaged accounts with penalties for early withdrawals, you can use HSA funds anytime for qualified healthcare costs. And if you're healthy now? The money keeps growing tax-free until you need it, whether that's next year or decades from now in retirement.

Pro Tip: Save every medical receipt, even for stuff you pay cash for now. Years down the road, you can pull that money out tax-free by showing those old receipts. Meanwhile, your crypto has been sitting there growing. My father-in-law just reimbursed himself for knee surgery from 2012 and basically got a 300% tax-free return.

Finding Crypto-Friendly HSA Providers

Not all HSA providers allow cryptocurrency investments. You'll need to look for providers that offer self-directed HSAs with brokerage capabilities. Companies like Lively, HealthSavings, and Optum Bank have started offering investment options beyond traditional mutual funds, though their crypto selections may vary.

Setting Up Your Crypto Tax Shelter: A Simple Guide

Setting up a tax-advantaged account for crypto isn't as complicated as it might seem.

Finding Your Crypto-Friendly Financial Partner

Not all financial institutions welcome crypto with open arms. You'll need to find specialized custodians that actually support digital assets in retirement accounts. Companies like iTrustCapital, BitIRA, and Equity Trust have built their services specifically for crypto investors like you.

Before signing up, check which cryptocurrencies they support. Some providers limit you to just Bitcoin and Ethereum, while others offer dozens of altcoins.

Security First, Always

Your crypto security matters even more inside tax-advantaged accounts since you can't just move assets in and out freely. You can't setup a cold wallet and move your crypto there, so you're relying the custodian to keep your investments safe from hackers. Ask potential custodians tough questions about how they protect your digital assets:

"Do you use multi-sig cold storage for most funds?"

"What insurance coverage do you provide?"

"Have you ever experienced a security breach?"

A good provider will proudly answer these questions with specific details, not vague reassurances.

Some custodians also restrict what you can do with your crypto. Many don't allow staking or yield farming inside retirement accounts. If passive income strategies are part of your plan, make sure to ask about these limitations upfront.

Getting Money Into Your Account

This part is straightforward. You have two main options:

1. Make a fresh contribution with cash from your bank account
2. Roll over funds from an existing IRA, 401(k), or HSA

Just remember: you can't directly move your existing crypto holdings into these accounts. The IRS requires you to contribute cash first, then buy crypto inside the account.

Global Crypto Tax Shelters: Beyond U.S. Borders

Not everyone lives in the United States, and crypto is global by nature. If you're investing from another country, you have your own set of tax tools. Let's take a quick trip around the world to see what's available.

United Kingdom: ISAs with a Twist

The UK's Individual Savings Accounts (ISAs) are similar to American Roth IRAs. They offer tax-free growth on investments. The catch? Standard ISAs don't directly support cryptocurrency. However, some clever investors use crypto ETFs or companies with significant blockchain exposure inside their ISAs to gain indirect crypto exposure while keeping the tax benefits.

Some UK investors also use Self-Invested Personal Pensions (SIPPs) to gain exposure to crypto through regulated investment vehicles. While direct crypto holdings aren't allowed, you can invest in crypto-focused funds that fit within pension regulations.

Australia: Super Solutions for Crypto

Down under, Australians have their Superannuation ("Super") retirement system. Traditional Super funds are slowly warming up to digital assets, with some now allocating small portions to crypto. For more control, self-managed super funds (SMSFs) give Australians direct access to cryptocurrency investments while maintaining tax advantages.

Canada: TFSAs and RRSPs

Canadians have two main tax-advantaged options: Tax-Free Savings Accounts (TFSAs) and Registered Retirement Savings Plans (RRSPs). Similar to the U.K., direct crypto investments aren't allowed in standard accounts, but Canadians can invest in regulated crypto ETFs that trade on the Toronto Stock Exchange.

This gives Canadian investors a unique advantage. They can gain exposure to Bitcoin and Ethereum through approved ETFs inside their tax-advantaged accounts without needing specialized custodians.

Going Beyond Traditional Structures

Some countries offer unique structures worth exploring:

- Singapore's Supplementary Retirement Scheme (SRS) provides tax advantages and more investment flexibility than many Western retirement accounts
- Malta and Portugal have developed crypto-friendly tax policies that make them popular with digital asset investors
- Switzerland's Pillar 3a retirement system has started allowing limited crypto exposure

Remember, tax laws change frequently, especially around cryptocurrency. What works today might not work tomorrow. Always check with a local tax professional before making major moves.

If you aren't ready to jump in the tax-advantaged boat yet, there are still things you can do to minimize your tax bill next year.

Tax Loss Harvesting: Making Lemons into Lemonade

The reality is that not every crypto purchase turns into a winner. But here's a secret that savvy investors know: even your losing investments can be valuable at tax time.

Tax loss harvesting is a strategy that turns your crypto "failures" into tax breaks. When your investments drop in value, you can sell them to lock in those losses on paper. Then, you use those losses to offset gains from your winning investments, slashing your tax bill in the process.

The Crypto Tax Loophole That Won't Last Forever

Here's where crypto has a unique advantage over stocks: the "wash sale" loophole. With stocks, if you sell at a loss, you have to wait 30 days before buying back the same asset, or the IRS won't let you claim the loss. But this rule doesn't apply to crypto (at least not yet).

This means you can sell your Bitcoin at a loss today, claim the tax deduction, and buy back the exact same amount immediately. You keep your position while banking the tax benefit. The IRS is likely to close this loophole eventually, but for now, it's completely legal.

Smart Ways to Use Tax Loss Harvesting

Remember these key points when harvesting losses:

Your short-term losses are most valuable against short-term gains, which are taxed at higher rates. If you have both types of gains, offset the short-term ones first.

Keep a running tally of your crypto performance throughout the year. December isn't the only time to harvest losses. Market dips can happen anytime.

Document everything thoroughly. Record the exact time, price, and amount of each transaction. If the IRS ever questions your strategy, proper documentation makes all the difference.

And if your losses exceed your gains? You can deduct up to $3,000 against your regular income. Any remaining losses roll forward to future tax years, giving you tax benefits for years to come.

Taxes aren't the only thing to plan for. Responsible investing also means thinking ahead. That includes what happens to your crypto assets when you're no longer here.

Crypto Estate Planning: What Happens to Your Bitcoin When You're Gone?

Most crypto investors spend hours researching coins, setting up wallets, and tracking market movements. But few ever stop to consider a crucial question: What happens to your crypto when you're no longer around?

Unlike traditional assets that automatically transfer through banks and brokerages, crypto can vanish forever if you don't plan ahead. Your family can't access what they don't have the keys to. They might not ever even know it exists.

The $190 Million Cautionary Tale

In 2018, Gerald Cotten, the CEO of QuadrigaCX exchange, died unexpectedly during a trip to India. The problem? He was the only person with access to the exchange's wallets. The result? About $190 million in customer funds became permanently inaccessible.

True story. It's a warning for every crypto holder. Without proper planning, your digital assets could be lost forever, regardless of their value. Can you imagine being one of those customers?

If you don't want that to happen to you, you need a plan.

Creating Your Crypto Legacy Plan

Start with a simple crypto inventory. List everything you own and exactly how to access it:

- Which wallets do you use? Where are they stored?
- Do you have exchange accounts? What are they?
- Are you staking coins or using DeFi platforms?
- Where are your seed phrases and private keys stored?

This inventory should be thorough but extremely secure. Never put all this information in a single unencrypted document.

For security that doesn't sacrifice accessibility, consider breaking your plan into pieces:

- Put wallet addresses in your will or trust documents (these are public anyway).
- Store private keys or seed phrases in a separate secure location.
- Leave instructions on how these pieces fit together.

Many investors now use "dead man's switch" setups, where trusted family members get access to keys if you don't check in for a certain period. Others use multi-sig wallets that require multiple people to approve transactions.

You also don't have to wait until you're gone to transfer crypto if that's how you want to approach it.

Tax-Free Crypto Gifting: A Strategy for the Living

Each year, you can gift up to $19,000 per person (in 2025) without triggering gift taxes. This strategy shrinks your taxable estate while sharing wealth with loved ones during your lifetime.

The best part? If you gift crypto that later skyrockets in value, all that growth happens outside your estate. Your $19,000 gift of Ethereum today could be worth many times that in your recipient's hands years later, with no estate tax implications for you.

By combining thoughtful gifting with secure inheritance planning, you ensure your crypto wealth benefits those you care about most.

Common Questions About Tax-Advantaged Crypto Investing

What happens if I withdraw crypto from my IRA before retirement age? Early withdrawals (before age 59½) typically trigger income taxes plus a 10% penalty. For Roth IRAs, you can withdraw contributions (but not earnings) penalty-free after five years. You would pay income taxes at your normal income tax rate.

Can I stake my crypto inside a tax-advantaged account? Most crypto IRA custodians don't currently support staking due to technical and regulatory challenges. However, this is changing as providers develop more capabilities. Ask potential custodians specifically about staking if this is important to your strategy.

If I use a self-directed IRA LLC, can I manage my own crypto keys? Yes, with the right structure. Some investors use a "checkbook IRA" with an LLC to gain more direct control over their crypto assets. This approach requires careful setup and compliance to avoid prohibited transactions that could disqualify your entire IRA.

What's the biggest mistake people make with tax-advantaged crypto investing? Not starting early enough. The power of these accounts comes from long-term tax-free or tax-deferred growth. Even small contributions can grow substantially over decades, especially with volatile assets like cryptocurrency.

With that, now we're getting into the more advanced stuff. Let's continue down this path into the next chapter and explore some advanced investing techniques.

11

ADVANCED INVESTING TECHNIQUES: TAKING YOUR INVESTMENTS TO THE NEXT LEVEL

READY TO GO BEYOND JUST BUYING AND HOLDING CRYPTO? THERE ARE OTHER techniques that can truly amplify your returns while keeping risk in check. This chapter breaks down the strategies professional crypto investors use but rarely discuss in public forums: systematically rebalancing your portfolio to lock in gains, strategically using leverage when odds are in your favor, spotting promising new cryptocurrencies before they explode in value, and putting your assets to work generating passive income through smart contracts. These are practical approaches that can transform how you invest in digital assets.

Portfolio Rebalancing

Your crypto portfolio is kinda like a plant. It grows unevenly and needs regular pruning to stay healthy. That's where rebalancing comes in. At its core, it's about maintaining discipline when emotions want to take over. You sell some of your winners (yes, the ones still pumping) and buy more of your laggards. Sounds backward, right? But this counter-intuitive approach is exactly what keeps your risk levels in check and prevents a single skyrocketing asset from dominating your portfolio.

Why Rebalancing Matters

Without rebalancing, you might wake up to find that your carefully diversified portfolio has morphed into a high-risk gamble on a single token. I've seen investors who started with a balanced approach end up with 70% of their portfolio in a single asset just because it performed well. When that asset crashed, it wiped out years of gains.

- **Risk Control:** Prevents overexposure to high-risk assets by redistributing gains into safer investments.
- **Profit Lock-In:** Ensures that profits from surging assets are realized before a potential downturn.
- **Avoiding Market Trends:** Encourages disciplined investing rather than reacting to emotional swings.
- **Maintaining Strategy Alignment:** Keeps the portfolio consistent with long-term financial goals.

Over time, rebalancing smooths out extreme instability so investors can manage downside risks while still benefiting from market uptrends.

Setting a Rebalancing Strategy That Works

Don't become a rebalancing maniac, checking prices hourly and tweaking allocations. That road also leads to stress and overtrading. It ends up being counterproductive. Instead, pick a rebalancing strategy you can stick with.

Most successful crypto investors use one of these approaches:

1. Threshold-Based Rebalancing: You set tolerance bands for each asset (say, ±5% from your target). When Bitcoin grows from your target 40% to over 45% of your portfolio, you sell some and redistribute. Simple but effective.

2. Time-Based Rebalancing: Mark your calendar for monthly, quarterly, or annual portfolio check-ups. This prevents the obsessive chart-watching that drives many crypto investors crazy. On your designated day, you adjust everything back to your target percentages, regardless of market conditions.

3. Algorithmic & Automated Rebalancing: Let technology do the heavy lifting. Platforms like Shrimpy automatically rebalance based on your rules. 3Commas offers bot-driven solutions for keeping your allocations in check.

No single approach works for everyone. A day trader might prefer threshold-based rebalancing, while someone with a full-time job outside crypto might opt for quarterly rebalancing or automation.

Practical Rebalancing Scenarios

Let me walk you through some real-world examples of how rebalancing plays out:

1. The Bitcoin Breakout Scenario

You started with 40% Bitcoin, 30% Ethereum, and 30% stablecoins. Bitcoin goes on a tear and suddenly represents 60% of your portfolio. Your gut screams "let it ride!" but your rebalancing strategy says otherwise. You sell some Bitcoin, distributing the proceeds to ETH and stablecoins to restore your original allocation. A month later, when Bitcoin pulls back 20% (as it often does even in bull markets), you've already banked those gains.

2. The Sector Rotation Play

Your portfolio holds mostly Layer 1 blockchains (Ethereum, Solana, Avalanche), but you notice DeFi tokens outperforming them. Instead of completely abandoning your strategy, you rebalance by shifting a portion of your Layer 1 holdings into select DeFi projects. This controlled approach gives you exposure to the hot sector without going all-in on a trend that might reverse.

3. The Risk Reduction Reality Check

You start with 50% altcoins and 50% Bitcoin/stablecoins. The altcoin rally pushes this allocation to 70% altcoins. Excitement builds, but so does risk. By rebalancing back to your original 50/50 split, you lock in altcoin profits while ensuring a major correction won't devastate your portfolio. Your friends might call you crazy for selling during a rally, but you'll thank yourself during the inevitable downturn.

It's easy to assume that an investment that's been pumping will continue its upward trajectory forever. Take your profits while you can! It leads to better long-term results.

Tax Considerations When Rebalancing

Here's the fun part again: taxes. Every time you rebalance by selling assets, you potentially trigger taxable events. The impact varies dramatically depending on where you live.

A few critical points to remember:

- Know your jurisdiction's rules on short vs. long-term capital gains.
- Keep meticulous records of your cost basis (what you paid) for each crypto purchase.
- Consider tax-advantaged accounts.
- Look into tax-loss harvesting.

Utilizing Smart Contract Platforms for Passive Income

Let's break down the three main strategies that can generate consistent returns on smart contract platforms without requiring you to obsessively check price charts.

Staking: The Digital Equivalent of Interest-Bearing Savings

Staking is the entry-level passive income strategy in crypto. You lock up tokens to help secure a blockchain network and earn rewards for your service.

Here's how it works: Proof-of-Stake blockchains need token holders to validate transactions. By committing your tokens, you essentially become a mini bank that processes network activity. The network pays you for this service with additional tokens—typically 4-15% annually depending on the blockchain.

Popular staking options include:

- Ethereum: Currently paying around 4-5% APY
- Cosmos: Offering approximately 8-12% APY
- Solana: Providing roughly 6-11% APY
- Polkadot: Generating about 10-14% APY

The beauty of staking is its simplicity. Stake 10 ETH earning 5% annually, and you'll have 10.5 ETH after year one and 11.025 after year two, compounding without any additional effort on your part.

The catch? Your tokens are often locked for specific periods. If prices crash while your assets are staked, you might be forced to watch your value evaporate without the ability to sell.

You might be thinking, "Why would I stake my crypto if I don't want to be a validator?" The short answer is you don't have to. Many popular centralized exchanges let you stake your crypto without running any hardware or software. They combine your tokens with others in a staking pool and either operate their own validators or delegate to trusted ones. As the validators earn rewards for securing the network, the exchange shares a portion of those rewards with you. It is often paid out like interest.

Liquidity Pools: Banking for the Crypto Economy

Decentralized exchanges use liquidity pools or collections of paired assets that enable trading. By adding your tokens to these pools, you earn a cut of every trade that uses your liquidity.

For example, if you add equal values of ETH and USDC to a Uniswap pool, traders swapping between these assets will pay you a portion of their trading fees. Active pools on major DEXs can generate anywhere from 5% to 30%+ annual returns, depending on trading volume.

Major platforms for liquidity provision include:

- Uniswap on Ethereum
- PancakeSwap on BNB Chain
- Curve Finance for stablecoin pairs
- Balancer for multi-asset pools

The real appeal? Unlike traditional finance where banks pocket the spread between deposit rates and lending rates, in DeFi, you capture that value directly. You become the bank.

Yield Farming: Advanced Multi-Layer Strategies

If staking is a savings account and liquidity pools are like being a banker, yield farming is like running a hedge fund. It involves strategically moving assets between different protocols to maximize returns from multiple sources simultaneously.

A basic yield farming strategy might look like this:

1. Deposit stablecoins into Aave as collateral
2. Borrow another asset against your collateral
3. Provide the borrowed asset to a liquidity pool earning high rewards
4. Collect liquidity fees plus additional token incentives

When executed correctly, yield farming can generate eye-popping APYs. Sometimes in the triple digits. But these returns come with compound risk and often require frequent portfolio adjustments.

The passive income strategies above sound amazing on paper. But they involve risks that YouTube influencers conveniently ignore when shilling their favorite platforms.

The Smart Contract Vulnerability Problem

Don't forget our previous discussion on the challenges with smart contracts. DeFi protocols run on code, and code can have bugs. In 2022 alone, hackers stole over $3 billion from DeFi protocols through various exploits. When smart contracts fail, they often fail catastrophically—with users losing 100% of their deposited funds.

Even audited projects aren't immune. Multiple "safe" and "audited" proto-cols have been drained through novel attack vectors. Your best protection? Diversification across platforms and keeping up with security practices.

Understanding Impermanent Loss

This counterintuitive concept catches many liquidity providers by surprise. If you deposit an ETH/USDC pair and ETH doubles in price, you'd have been better off simply holding ETH instead of providing liquidity.

Why? Because as ETH rises, the pool automatically sells some of your ETH for USDC to maintain balance. When you withdraw, you'll have more USDC but less ETH than you started with and miss out on some of that ETH price appreciation.

The more volatile the asset pair and the bigger the price movement, the greater your impermanent loss. Stablecoin pairs (like USDC/DAI) minimize this risk but offer lower returns.

The Yield Mirage

Many projects temporarily offer unsustainable yields to attract liquidity. A new DEX might offer 300% APY through token incentives, but these rates inevitably crash as more capital floods in or token prices fall.

Ask yourself: Where are these yields actually coming from? If the answer is "just newly minted tokens," be skeptical about long-term sustainability.

Regulatory Uncertainty

Governments worldwide are increasingly scrutinizing DeFi. A regulatory crackdown could dramatically impact certain protocols or strategies. Stay informed about developments in your jurisdiction and consider the regulatory risk profile of different platforms.

A Practical Example: Building a Passive Income Portfolio

Let's see how this might work in practice. Imagine you have $10,000 to allocate toward passive income strategies:

Conservative Approach ($10,000):

- $5,000 in Ethereum staking (5% APY)
- $3,000 in USDC/DAI liquidity on Curve (3-8% APY)
- $2,000 in Bitcoin on a lending platform like Celsius (3-6% APY)
- Expected Annual Return: $450-700 (4.5-7%)
- Risk Level: Moderate

Balanced Approach ($10,000):

- $3,000 in Ethereum staking (5% APY)
- $3,000 in ETH/USDC liquidity on Uniswap (10-20% APY)

- $2,000 in stablecoin lending on Aave (3-10% APY)$2,000 in mid-cap token staking (Cosmos, Polkadot, etc. at 10-14% APY)
- Expected Annual Return: $750-1,400 (7.5-14%)
- Risk Level: Medium-High

Aggressive Approach ($10,000):

- $2,000 in Ethereum staking (5% APY)
- $3,000 in liquidity for promising new DeFi protocols (20-80% APY)
- $3,000 in active yield farming strategies (15-100% APY)
- $2,000 in governance token staking for emerging protocols (10-30% APY)
- Expected Annual Return: $1,300-4,500 (13-45%)
- Risk Level: Very High

These are simplified examples. Actual implementation would require deeper protocol research and regular monitoring. The higher the potential return, the more hands-on management required.

The Sustainability Question

Some critics argue that current DeFi yields are artificially inflated and will inevitably decline as the market matures. They have a point. Many yields come from token incentives that can't last forever.

The counterargument? Traditional finance extracts enormous value through middlemen, fees, and opaque structures. DeFi redistributes that value to participants. While 100%+ APYs will likely fade, the efficiency of decentralized finance could sustain yields significantly higher than traditional investments over the long term.

Tips for Getting Started

If you're intrigued by the passive income potential of smart contracts, here's how to dip your toes without diving headfirst into the deep end:

1. Start with simple staking on major platforms like Ethereum or Cosmos

2. Use established protocols with long track records before newer platforms
3. Never commit funds you might need in an emergency
4. Split your capital across multiple strategies and platforms
5. Use tools like DeBank or Zapper to track your positions across protocols
6. Set calendar reminders to check performance and compound rewards

Remember, the tortoise often beats the hare. Consistent 8-12% returns with minimal risk can outperform chasing 100%+ yields that might disappear overnight or just result in total loss.

The most valuable yields are the ones you actually get to keep.

Earning passive income through smart contracts is a powerful way to put your assets to work. But crypto never stands still. While your money compounds in the background, your attention should stay on what's coming next. The real advantage comes from spotting market shifts and identifying emerging projects before they go mainstream. Let's talk about how to prepare for the next evolution.

Preparing for the Next Market Evolution

Tomorrow's revolution is already brewing in GitHub repositories and developer Discord channels. The investors who spot these shifts early don't just make profits, they sidestep catastrophic losses when outdated technologies collapse.

Think about those who recognized DeFi's potential in early 2020, before the summer explosion. Or NFT collectors who bought digital art before celebrities jumped in. Or those who moved to stablecoins before the 2022 crash. Being ahead of market transitions isn't just profitable. It's protective.

Reading the Market's Mood Swings

Crypto market sentiment shifts like weather patterns—sometimes gradually, sometimes violently. These psychological currents drive capital flows between different sectors and approaches:

During bull markets, money flows from Bitcoin to large-caps to mid-caps to small-caps and finally to the riskiest microcap projects. Fear of missing out pushes investors to chase increasingly speculative opportunities.

During bear markets, this process reverses. Capital rushes back to safety from microcaps to established projects to Bitcoin and finally to stablecoins. Fear of losing everything trumps the fear of missing out.

These are mass psychological shifts that show up as price shifts. And they repeat with surprising regularity. Recognizing where we are in this cycle gives you an enormous advantage.

The institutional money entering crypto adds another dimension. Traditional finance players typically start with Bitcoin exposure, then venture into Ethereum, before exploring DeFi and other sectors. Their entry points often signal the legitimization of specific crypto sectors.

Here's how to develop your trend-spotting capability:

Follow the Developers, Not the Influencers

The next big thing in crypto is being built right now by people coding, not those making YouTube thumbnails with shocked faces. Track GitHub commits, not subscriber counts.

Monitor technical discussions on crypto Twitter/X, Discord servers, and developer forums. Pay attention when multiple respected builders start exploring the same technology concept.

Watch Smart Money Movements

Track where venture capital firms like Paradigm, a16z, and Multicoin Capital are investing. Their research teams spend millions identifying promising sectors before they go mainstream.

Wallet tracking tools can show you what veteran crypto investors are buying before major price movements. Services like Nansen and Santiment offer insights into whale behavior that often precedes market shifts.

Identify the Pain Points

The most successful crypto innovations solve real problems. Ethereum addressed Bitcoin's limited programmability. Uniswap solved centralized exchange limitations. Layer 2 solutions tackled Ethereum's high fees.

Ask yourself: What's the biggest unsolved problem in crypto right now? High fees, cross-chain interoperability, user experience, regulatory compliance, scaling? The projects tackling these problems effectively will lead the next cycle.

Market transitions don't announce themselves with convenient warning signs. They happen while most investors are distracted by current opportunities. Preparation is everything.

Spotting a trend or opportunity early is only half of the equation. What you do with that insight is what separates speculators from strategic investors. Once you identify where the market is heading, the next step is positioning your portfolio to capitalize on emerging projects without overexposing yourself to unnecessary risk. Here's how to approach it.

From Theory to Practice: Building Your Exposure

How much of your portfolio should venture into emerging cryptocurrencies? That depends on your risk tolerance, but here's a framework:

Conservative Approach (5-10% allocation):

- Focus on established alternatives to Ethereum (Solana, Avalanche, Polkadot)
- Limit individual project exposure to 1-2% of your total portfolio
- Only invest in projects with working products and significant adoption

Moderate Approach (10-20% allocation):

- Include both established and promising mid-cap projects
- Consider sector-based diversification (some DeFi, some infrastructure, some NFT/gaming)
- Allow 2-3% allocation to individual high-conviction projects

Aggressive Approach (20-30% allocation):

- Include early-stage projects with strong fundamentals but limited current adoption
- Consider small positions (0.5-1%) in very early projects with exceptional teams
- Actively rebalance as winners emerge, taking profits regularly

This filtering process eliminates 95% of new projects (as it should). Quality investments are rare, especially in emerging cryptocurrencies.

Whatever you do, don't commit too much of your portfolio these emerging projects. Too many investors do it. They put 80% in speculative microcaps and 20% in established projects. Even the most brilliant innovation needs time to prove itself, and many conceptually sound projects fail in execution.

And I know you have burning questions about meme coins like DOGE and PEPE. Elon Musk summed it up perfectly in a recent Forbes article: "If you expect to win at meme coins you're being foolish, you're not going to win, don't sink your life savings into a memecoin ... don't bet the farm." Meme coins generally lack utility beyond speculation. Many experience rapid pumps followed by steep declines. Some never recover.

The Hidden Dangers Most Investors Ignore

Beyond the obvious risk of picking the wrong project, emerging cryptocurrencies carry additional dangers that can trap even experienced investors:

The Liquidity Trap: Thinly traded tokens can experience significant slippage—sometimes 10–30% or more—especially during volatile periods or large sell-offs. Always check 24-hour trading volume before investing significant amounts.

The Regulatory Roulette: New projects operate in regulatory gray areas. A single SEC announcement or country ban can send prices plummeting overnight. Those cute animal tokens? Many could be classified as unregistered securities.

The Abandoned Project Reality: Development teams lose funding, motivation, or unity. Once-promising projects get abandoned with vague announcements about "exploring strategic alternatives." Always have an exit strategy.

The Exchange Delisting Risk: Smaller tokens can get delisted from exchanges due to low volume or regulatory concerns, leaving investors with assets they can't easily sell.

Some investors argue you're better off just sticking with Bitcoin and Ethereum. Their rationale? The risk-adjusted returns from carefully selected blue-chips often outperform a portfolio of speculative bets after accounting for the inevitable losers.

The Reality Check

Investing in emerging cryptocurrencies isn't a get-rich-quick scheme. It's like investing in a startup. Things need time to bake and most projects won't even survive a full market cycle. The ones that do can deliver extraordinary returns, but they're the exception, not the rule.

Your edge comes from better research, stronger conviction, and stricter discipline than the average investor. Set clear criteria for both buying and selling. Know what success looks like beforehand, and take profits when projects meet your targets. This discipline separates investors who capture value from those who watch paper gains evaporate.

Remember: In emerging crypto, your goal isn't to catch every winner—it's to find a few genuine innovations while avoiding the sea of worthless tokens that will eventually go to zero.

Taking a measured approach prevents both missing major movements and over-committing to false starts.

Maintaining Dry Powder

The best opportunities often appear during market distress when quality projects drop 80-90% from their highs. Without available capital, you'll miss these generational buying moments.

Keep 10-20% of your crypto portfolio in stablecoins, regardless of market conditions. You need to be prepared with ammunition when rare, asymmetric opportunities emerge. You don't want to get caught with empty pockets when something comes around that you think could be the next Ethereum.

During the darkest days of crypto winter, when despair is maximum and prices seem destined to fall forever, deploy this capital methodically into high-conviction, cash-flow positive projects with strong fundamentals.

The Final Word on Market Evolution

Preparing for the next market shift is about awareness and flexibility. The best crypto investors combine:

- Deep research to identify emerging trends
- Disciplined position sizing to manage risk
- Strategic diversification across market sectors
- Patient capital deployment during extreme conditions
- Emotional fortitude to act counter to market sentiment

Remember this: By the time a crypto trend is obvious to everyone, the easiest money has already been made. The biggest returns come from seeing what others don't yet.

Don't chase yesterday's momentum. Position for tomorrow's paradigm shift.

CONCLUSION

We've traveled a long road together. From blockchain basics to tax strategies, you now grasp what most investors never bother to learn. But knowledge without action is just trivia.

So what happens next?

Take a moment and imagine yourself a year from now. You've set up your first crypto wallet. You've made your initial investment. You've weathered your first market dip without panic-selling (maybe even bought more during the drop).

How would that feel?

Maybe you're thinking, "I could finally feel like I'm getting caught up on saving for retirement."

Maybe it's, "I could build wealth on my own terms, outside the traditional system that wasn't designed for people like me."

Or simply, "I could stop being afraid of what I don't know."

Your crypto journey might start small. $50 a month, perhaps. But what about five years from now? Ten? The compound effect of consistent investing coupled with the growth potential of blockchain technology could transform your financial reality.

André De Shields, a Broadway performer and Tony Award winner once said, "slowly is the fastest way to get to where you want to be."

Picture yourself reaching milestones you once thought impossible:

- Paying off lingering debt with investment returns
- Building a college fund that actually keeps pace with rising tuition
- Creating passive income streams that reduce your dependence on a paycheck
- Funding the dream vacation, home renovation, or early retirement you've been postponing

This isn't fantasy. Real people like you have done it. How many Bitcoin millionaires are out there? People who put $50 down on a whim and now never have to work a day in their lives again? Those opportunities may still exist.

What could this mean for you?

The opportunity cost of inaction is real. While others build wealth in this new financial system, will you still be on the outside looking in? Will you still be telling yourself "it's too late" or "it's too risky" five years from now?

I can't promise you'll get rich. No honest crypto book should. But I can promise that taking informed action based on what you've learned creates possibilities that sitting still never will.

Start today. Begin small if needed. Make mistakes early when the stakes are low.

And remember: you don't need to do this alone. Join online communities where beginners support each other. Follow credible voices who share insights without overembellishing. Revisit this book when you feel unsure.

The future of finance is being written right now, and you hold the pen. What story will you tell?

Your first transaction awaits. Your first market cycle beckons. Your financial transformation starts now.

All that's left is for you to begin.

Disclaimer:

The information in this book is for educational purposes only and should not be considered financial or investment advice. Cryptocurrency markets are highly volatile, and all investments carry inherent risks. Readers should conduct their own research and consult a qualified financial professional before making any investment decisions. The author and publisher are not responsible for any financial losses incurred as a result of using the information provided. Past performance does not guarantee future results, and investing always involves uncertainty.

REFERENCES

- Authy. "Protect Your Crypto Accounts with Authy 2FA." https://authy.com/blog/protecting-your-cryptocurrency-with-authy/
- Axie Infinity Whitepaper. https://whitepaper.axieinfinity.com/
- Binance. "Binance DEX Overview." https://docs.binance.org/
- Binance. "Support Announcement: Binance Updates." https://www.binance.com/en/support/announcement/detail/360028031711
- Binance Academy. "Proof of Work Explained." https://academy.binance.com/en/articles/proof-of-work-explained
- BitGo. "Wallet and Custody Guide." https://www.bitgo.com/resources/wallet-and-custody-guide/
- Bitconned. Directed by Bryan Storkel. Netflix, 2024.
- Bitcoinist. "Solana (SOL) Jumped from $1.50 to $259 ATH in 2021." Bitcoinist, 2024. https://bitcoinist.com/solana-sol-jumped-from-1-50-to-259-ath-in-2021-experts-reckon-this-under-0-10-altcoin-could-replicate-the-same/
- Blockchain.com. Genesis Block Explorer. https://www.blockchain.com/explorer/blocks/btc/0
- Bloomberg. "JPMorgan's Blockchain Division Processing $300 Billion Daily." https://www.bloomberg.com/news/articles/2023-09-12/jpmorgan-s-blockchain-division-now-moving-300-billion-daily
- Bloomberg News. "El Salvador President Says Nation Adopts Bitcoin as Legal Tender." Bloomberg, June 9, 2021. https://www.bloomberg.com/news/articles/2021-06-09/el-salvador-president-says-nation-adopts-bitcoin-as-legal-tender
- Basic Attention Token. "Basic Attention Token (BAT)." https://basicattentiontoken.org/
- Cambridge Bitcoin Electricity Consumption Index. https://ccaf.io/cbeci/index
- CarbonCredits.com. "The Energy Debate: How Bitcoin Mining, Blockchain, and Cryptocurrency Shape Our Carbon Future." Last modified 2024. https://carboncredits.com/the-energy-debate-how-bitcoin-mining-blockchain-and-cryptocurrency-shape-our-carbon-future
- Chainalysis. "2024 Crypto Hacking Year in Review: $1.7B Stolen, a Drop from 2022's Record-Breaking $3.7B." *Chainalysis Blog*, February 7, 2024. https://www.chainalysis.com/blog/crypto-hacking-stolen-funds-2024
- Chainlink Docs. "Decentralized Oracle Networks." https://docs.chain.link/docs/
- Chainlink Labs. "What Is Chainlink?" https://chain.link/education
- Circle. "What Is USDC?" https://www.circle.com/en/usdc

REFERENCES

- CNN Business. "Coincheck Cryptocurrency Exchange Hack in Japan." CNN, January 29, 2018. https://money.cnn.com/2018/01/29/technology/coincheck-cryptocurrency-exchange-hack-japan/index.html
- Coin Bureau. "Cryptocurrency Margin Trading: Everything You Need to Know." https://coinbureau.com/education/cryptocurrency-margin-trading
- CoinDesk. "The Genesis Block: The First Bitcoin Block." https://www.coindesk.com/learn/the-genesis-block
- CoinDesk. "Vietnam Investigates ICO Fraud After $660 Million in Losses." CoinDesk, April 11, 2018. https://www.coindesk.com/markets/2018/04/11/vietnam-investigates-ico-fraud-after-660-million-in-losses-reported
- Coinbase. "Coinbase Wallet Security." https://www.coinbase.com/security/wallet-security
- Coinbase. "What Is a Crypto Wallet?"
- Coinbase. "What Are Ethereum Layer 2 Blockchains and How Do They Work?" https://www.coinbase.com/learn/crypto-basics/what-are-ethereum-layer-2-blockchains-and-how-do-they-work
- CoinLedger. "Best Crypto Tax Software in 2024." https://coinledger.io/tools/best-crypto-tax-software
- CoinLedger. "IRS Crypto Cost Basis Reporting Requirements (Rev. Proc. 2024-28)." https://coinledger.io/blog/irs-new-crypto-cost-basis-rules-rev-proc-2024-28
- CoinMarketCap. "Bitcoin Halving." https://coinmarketcap.com/halving/bitcoin/
- CoinMarketCap. "Cryptocurrency Market Capitalizations." https://coinmarketcap.com/charts/
- CoinMarketCap. "Top Cryptocurrency Exchanges by Volume." https://coinmarketcap.com/rankings/exchanges/
- CoinMarketCap. "Top DeFi Tokens by Market Capitalization." https://coinmarketcap.com/view/defi/
- CoinMarketCap. "Technical Analysis/Trend Analysis (TA)." https://coinmarketcap.com/academy/glossary/technical-analysis-trend-analysis-ta
- CoinTelegraph. "Top 5 Ways to Earn Passive Income with Crypto in 2025." https://cointelegraph.com/learn/articles/ways-to-earn-passive-income-with-crypto
- Cointelegraph. "The Mt. Gox Bitcoin Heist, and Why It Still Matters." https://cointelegraph.com/learn/articles/the-mt-gox-bitcoin-heist
- Cosmos Network. "Cosmos Whitepaper." https://cosmos.network/whitepaper
- CryptoSlate. "Satis Group Report: 78% of ICOs Are Scams." https://cryptoslate.com/satis-group-report-78-of-icos-are-scams/
- Decentraland Docs. "What Is Decentraland?" https://docs.decentraland.org/
- dYdX Foundation. "What Is dYdX?" https://dydx.foundation/
- Electric Coin Company. "Zcash: Privacy-Protecting Digital Currency." https://z.cash
- Estonian Government eHealth Foundation. "Blockchain and Healthcare." https://e-estonia.com/blockchain-healthcare-estonian-experience/

- ETF Market Insights. "Benefits of a Core-Satellite Investment Approach." https://etfmarketinsights.com/blog/benefits-of-a-core-satellite-investment-approach/
- Ethereum Foundation. "The Merge." https://ethereum.org/en/upgrades/merge/
- Ethereum.org. "Ethereum Whitepaper." https://ethereum.org/en/whitepaper/
- Fear and Greed Index. https://alternative.me/crypto/fear-and-greed-index/
- Federal Bureau of Investigation (FBI) Internet Crime Complaint Center (IC3). "Public Service Announcement: Increase in Crypto-Related Scams Targeting Americans." https://www.ic3.gov/PSA/2025/PSA250226
- Federal Reserve History. "The Great Recession and Its Aftermath." https://www.federalreservehistory.org/essays/great-recession-and-its-aftermath
- Filecoin Foundation. "What Is Filecoin?" https://filecoin.io
- Forbes. "What Crypto Investors Should Know About the IRS Tax Regime for 2025." https://www.forbes.com/sites/digital-assets/2024/07/07/what-crypto-investors-should-know-about-the-irs-tax-regime-for-2025
- Forbes. "What Is DeFi?" https://www.forbes.com/sites/digital-assets/article/what-is-defi/
- Gemini. "User Insurance and Security Practices." https://www.gemini.com/security
- Genesis Block - farawaystars. https://farawaystars.com/bitcoin-genesis-block/
- Harvard Business Review. "The Truth About Blockchain." https://hbr.org/2017/01/the-truth-about-blockchain
- IBM. "What Are Smart Contracts?" https://www.ibm.com/topics/smart-contracts
- IBM. "What Is Blockchain Technology?" https://www.ibm.com/topics/what-is-blockchain
- Independent. "Floyd Mayweather and DJ Khaled Fined Over Cryptocurrency ICO Scam." Independent, 2024. https://www.independent.co.uk/tech/floyd-mayweather-dj-khaled-cryptocurrency-ico-scam-bitcoin-centra-tech-sec-a8595971.html
- Investopedia. "Dollar-Cost Averaging (DCA) Explained." https://www.investopedia.com/terms/d/dollarcostaveraging.asp
- Investopedia. "Smart Contracts: What You Need to Know." https://www.investopedia.com/terms/s/smart-contracts.asp
- IRS. "401(k) Limit Increases to $23,500 for 2025, IRA Limit Remains $7,000." IRS, 2024. https://www.irs.gov/newsroom/401k-limit-increases-to-23500-for-2025-ira-limit-remains-7000
- IRS. "Digital Assets." Internal Revenue Service. https://www.irs.gov/filing/digital-assets
- IRS. "Treasury, IRS Issue Final Regulations Requiring Broker Reporting of Sales and Exchanges of Digital Assets." https://www.irs.gov/newsroom/treasury-irs-issue-final-regulations-requiring-broker-reporting-of-sales-and-exchanges-of-digital-assets-that-are-subject-to-tax-under-current-law-additional-guidance-to-provide-penalty-relief-address

REFERENCES

- JPMorgan Chase. "Kinexsys by J.P. Morgan." https://www.jpmorgan.com/kinexys/index
- Larva Labs. "CryptoPunks." https://www.larvalabs.com/cryptopunks
- Ledger. "Hot Wallet vs Cold Wallet: What's the Difference?" https://www.ledger.com/academy/topics/security/hot-wallet-vs-cold-crypto-wallet-whats-the-difference
- Litecoin Foundation. "About Litecoin." https://litecoin.org
- Los Angeles Times. "Mt. Gox Bankruptcy and Bitcoin Loss." LA Times, February 28, 2014. https://www.latimes.com/business/technology/la-fi-tn-mt-gox-bankruptcy-bitcoins-20140228-story.html
- MakerDAO. "How DAI Works." https://makerdao.com/en/whitepaper/
- Monero.How. "How Monero's Privacy Works." https://www.monero.how/how-does-monero-privacy-work
- Monero Project. "Monero: Private Digital Currency." https://www.getmonero.org/resources/moneropedia/
- Multi-Signature (Multi-Sig) Wallets in Cryptocurrency: Enhancing Bitcoin Security and Flexibility. https://www.tftc.io/multi-signature-multi-sig-wallets-in-cryptocurrency-enhancing-bitcoin-security-and-flexibility/
- Nakamoto, Satoshi. "Bitcoin: A Peer-to-Peer Electronic Cash System." https://bitcoin.org/bitcoin.pdf
- Nasdaq. "How Bitcoin Fear and Greed Index Trading Strategy Beats Buy-and-Hold Investing." https://www.nasdaq.com/articles/how-bitcoin-fear-and-greed-index-trading-strategy-beats-buy-and-hold-investing
- NBA Top Shot. https://nbatopshot.com/
- Netflix. *Biggest Heist Ever*. Documentary. Directed by Seth Porges. Released 2024. https://www.netflix.com
- Netflix. *Bitconned*. Documentary. Directed by Bryan Storkel. Released January 2024. https://www.netflix.com
- NinjaTrader. "Cryptocurrency Futures Trading Strategies." https://ninjatrader.com/futures/blogs/cryptocurrency-futures-trading-strategies
- NIST. "Digital Identity Guidelines." https://pages.nist.gov/800-63-3/
- OSL. "How to Rebalance a Cryptocurrency Portfolio Effectively." https://osl.com/academy/article/how-to-rebalance-a-cryptocurrency-portfolio-effectively
- PancakeSwap Docs. "Getting Started with PancakeSwap." https://docs.pancakeswap.finance/
- Polkadot Network. "Polkadot Wiki." https://wiki.polkadot.network/
- Protocol Labs. "Filecoin Documentation." https://docs.filecoin.io/
- RealT. "Tokenized Real Estate Investment Platform." https://realt.co/
- Reuters. "Crypto Hackers Stole Around $1.7 Billion in 2023." Reuters, January 24, 2024. https://www.reuters.com/technology/cybersecurity/crypto-hackers-stole-around-17-bln-2023-report-2024-01-24

- Shift Markets. "What Is a Hybrid Crypto Exchange?" ShiftMarkets.com. https://www.shiftmarkets.com/blog/what-is-a-hybrid-crypto-exchange
- Shrimpy. "Automated Portfolio Management for Crypto Investors." https://www.shrimpy.io
- Solana Foundation. "Solana Documentation: Proof of History." https://solana.com/docs
- Synthetix Docs. "What Are Synthetic Assets?" https://docs.synthetix.io/
- TechCrunch. "Bitconnect, Accused of Running a Ponzi Scheme, Shuts Down." TechCrunch, January 16, 2018. https://techcrunch.com/2018/01/16/bitconnect-which-has-been-accused-of-running-a-ponzi-scheme-shuts-down
- Tether. "Transparency and Reserves." https://tether.to/en/transparency/
- The Guardian. "QuadrigaCX Canada Cryptocurrency Exchange Locked Funds." The Guardian, February 4, 2019. https://www.theguardian.com/technology/2019/feb/04/quadrigacx-canada-cryptocurrency-exchange-locked-gerald-cotten
- The Nitty-gritty of Bitcoin Mining: How Exactly Does it Work? - UPay Blog. https://blog.upay.best/the-nitty-gritty-of-bitcoin-mining-how-exactly-does-it-work/
- The Sandbox. "Virtual Land Economy." https://www.sandbox.game/en/land/
- The Wall Street Journal. "Bitcoin Is Surging, and the IRS Is Watching Closer Than Ever." https://www.wsj.com/personal-finance/taxes/the-hardest-part-of-cashing-in-bitcoin-is-getting-the-taxes-right-88f85487
- THORChain Docs. "Cross-Chain Liquidity Protocol." https://docs.thorchain.org/
- TradingView. "Bitcoin Dominance Chart." https://www.tradingview.com/symbols/CRYPTOCAP-BTC.D/
- Trakx. "Bitcoin Pizza Day: The $680 Million Pizza Order That Made History." Trakx.io. https://trakx.io/resources/insights/bitcoin-pizza-day/#:~:text=Bitcoin%20Pizza%20Day%2C%20celebrated%20every,first%20step%20into%20everyday%20commerce.
- Trezor. "Trezor Hardware Wallet Models." https://trezor.io/compare
- Trust Wallet. "Trust Wallet: Secure Multi-Coin Wallet." https://trustwallet.com/
- U.S. Department of the Treasury. "Troubled Asset Relief Program (TARP)." https://home.treasury.gov/data/troubled-assets-relief-program
- Uniswap Docs. "What Is Uniswap?" https://docs.uniswap.org/
- Wallet Guard. "Protecting Web3 Users from Phishing and Fraud." https://walletguard.app/
- Washington Post News. "Bitcoin Mining: Understanding the Backbone of Bitcoin's Blockchain." https://wpostnews.com/bitcoin-mining-understanding-the-backbone-of-bitcoins-blockchain/
- Web3 Foundation. "Introduction to Polkadot." https://polkadot.network/Polkadot-whitepaper.pdf
- World Economic Forum. "The Future of Financial Infrastructure." https://www3.weforum.org/docs/WEF_The_future_of_financial_infrastructure.pdf

REFERENCES

- Yahoo Finance. "FTX's Collapse Wiped Out $200 Billion." Yahoo Finance, 2024. https://finance.yahoo.com/news/ftxs-collapse-wiped-200-billion-154835453.html
- Yearn Finance. "About Yearn." https://docs.yearn.finance/
- Zcash Docs. "zk-SNARKs Explained." https://z.cash/technology/zksnarks/
- 3Commas. "Smart Trading Terminals and Automated Bots." https://3commas.io

www.ingramcontent.com/pod-product-compliance
Lightning Source LLC
Chambersburg PA
CBHW071651210326
41597CB00017B/2179